Image Fusion in Remote Sensing

Conventional and Deep Learning Approaches

Synthesis Lectures on Image, Video, and Multimedia Processing

Editor
Alan C. Bovik, *University of Texas, Austin*

The Lectures on Image, Video and Multimedia Processing are intended to provide a unique and groundbreaking forum for the world's experts in the field to express their knowledge in unique and effective ways. It is our intention that the Series will contain Lectures of basic, intermediate, and advanced material depending on the topical matter and the authors' level of discourse. It is also intended that these Lectures depart from the usual dry textbook format and instead give the author the opportunity to speak more directly to the reader, and to unfold the subject matter from a more personal point of view. The success of this candid approach to technical writing will rest on our selection of exceptionally distinguished authors, who have been chosen for their noteworthy leadership in developing new ideas in image, video, and multimedia processing research, development, and education.

In terms of the subject matter for the series, there are few limitations that we will impose other than the Lectures be related to aspects of the imaging sciences that are relevant to furthering our understanding of the processes by which images, videos, and multimedia signals are formed, processed for various tasks, and perceived by human viewers. These categories are naturally quite broad, for two reasons: First, measuring, processing, and understanding perceptual signals involves broad categories of scientific inquiry, including optics, surface physics, visual psychophysics and neurophysiology, information theory, computer graphics, display and printing technology, artificial intelligence, neural networks, harmonic analysis, and so on. Secondly, the domain of application of these methods is limited only by the number of branches of science, engineering, and industry that utilize audio, visual, and other perceptual signals to convey information. We anticipate that the Lectures in this series will dramatically influence future thought on these subjects as the Twenty-First Century unfolds.

Virtual Reality and Virtual Environments in 10 Lectures
Stanislav Stanković
2015

Dictionary Learning in Visual Computing
Qiang Zhang and Baoxin Li
2015

Combating Bad Weather Part II: Fog Removal from Image and Video
Sudipta Mukhopadhyay and Abhishek Kumar Tripathi
2015

Combating Bad Weather Part I: Rain Removal from Video
Sudipta Mukhopadhyay and Abhishek Kumar Tripathi
2014

Image Understanding Using Sparse Representations
Jayaraman J. Thiagarajan, Karthikeyan Natesan Ramamurthy, Pavan Turaga, and Andreas Spanias
2014

Contextual Analysis of Videos
Myo Thida, How-lung Eng, Dorothy Monekosso, and Paolo Remagnino
2013

Wavelet Image Compression
William A. Pearlman
2013

Remote Sensing Image Processing
Gustavo Camps-Valls, Devis Tuia, Luis Gómez-Chova, Sandra Jiménez, and Jesús Malo
2011

The Structure and Properties of Color Spaces and the Representation of Color Images
Eric Dubois
2009

Biomedical Image Analysis: Segmentation
Scott T. Acton and Nilanjan Ray
2009

Joint Source-Channel Video Transmission
Fan Zhai and Aggelos Katsaggelos
2007

Super Resolution of Images and Video
Aggelos K. Katsaggelos, Rafael Molina, and Javier Mateos
2007

© Springer Nature Switzerland AG 2022, corrected publication 2023
Reprint of original edition © Morgan & Claypool 2021

Image Fusion in Remote Sensing: Conventional and Deep Learning Approaches

Arian Azarang and Nasser Kehtarnavaz

ISBN: 978-3-031-01128-3   paperback
ISBN: 978-3-031-02256-2   ebook
ISBN: 978-3-031-00217-5   hardcover

DOI 10.1007/978-3-031-02256-2

A Publication in the Springer series
SYNTHESIS LECTURES ON IMAGE, VIDEO, AND MULTIMEDIA PROCESSING

Lecture #13
Series Editor: Alan C. Bovik, *University of Texas, Austin*
Series ISSN
Print 1559-8136   Electronic 1559-8144

Image Fusion in Remote Sensing

Conventional and Deep Learning Approaches

Arian Azarang and Nasser Kehtarnavaz
The University of Texas at Dallas

*SYNTHESIS LECTURES ON IMAGE, VIDEO, AND MULTIMEDIA
PROCESSING #13*

ABSTRACT

Image fusion in remote sensing or pansharpening involves fusing spatial (panchromatic) and spectral (multispectral) images that are captured by different sensors on satellites. This book addresses image fusion approaches for remote sensing applications. Both conventional and deep learning approaches are covered. First, the conventional approaches to image fusion in remote sensing are discussed. These approaches include component substitution, multi-resolution, and model-based algorithms. Then, the recently developed deep learning approaches involving single-objective and multi-objective loss functions are discussed. Experimental results are provided comparing conventional and deep learning approaches in terms of both low-resolution and full-resolution objective metrics that are commonly used in remote sensing. The book is concluded by stating anticipated future trends in pansharpening or image fusion in remote sensing.

KEYWORDS

image fusion in remote sensing, pansharpening, deep learning-based image fusion, fusion of spatial and spectral satellite images

Contents

Preface

Images captured by Earth observation satellites (called remote sensing images) are utilized in many applications. For example, farming has been made more effective, or weather prediction has been made more reliable by using remote sensing images. There are different image sensors on satellites capturing different attributes of the Earth surface. This book is about fusing or combining these different types of images into more informative ones. Image fusion in remote sensing is used in applications such as land-cover mapping, spectral unmixing, target identification, and anomaly detection. Recently, deep learning approaches have been applied to remote sensing image fusion.

The idea of writing this book came about out of the discussions between the authors regarding Arian Azarang's doctoral dissertation topic. Essentially, this book involves a rewriting of the deep learning approaches developed and covered in Arian Azarang's doctoral dissertation in the format of a book. It is written in order to provide an understanding of the state-of-the-art in remote sensing image fusion by discussing both conventional and deep learning approaches. We hope this book would be beneficial to scientists and researchers working in the remote sensing area.

As a final note, the author Arian Azarang would like to dedicate this book to his parents, Habib Azarang and Shoaleh Asami, for their unconditional support and love.

Arian Azarang and Nasser Kehtarnavaz
Feburary 2021

The original version of this book has been revised: The authors listed in the table of contents was incorrectly presented in the initially published online version of this book. The correct authors should read as Arian Azarang and Nasser Kehtarnavaz. This has now been corrected. A correction to this book can be found at https://doi.org/10.1007/978-3-031-02256-2_8.

CHAPTER 1

Introduction

1.1 SCOPE

Image fusion in remote sensing involves fusing spectral and spatial information that are captured by different sensors. Fusion in remote sensing, also called Pansharpening [1], of spatial and spectral images are used in different applications such as land cover classification [2], spectral unmixing [3], change detection [4], and target detection [5]. Earth observation satellites are normally equipped with two sensors which capture PANchromatic (PAN) and MultiSpectral (MS) images. A PAN image reflects the spatial information of a land scene and a MS image reflects the spectral information of the same scene. This lecture series book first provides an overview of remote sensing and then covers the fusion techniques for combining PAN and MS images. The techniques discussed include both conventional and deep learning-based ones.

1.2 ORGANIZATION

The book is organized into the following chapters.

CHAPTER 2

This chapter discusses an overview of remote sensing concepts. First, the definitions of spatial, spectral, radiometric, and temporal resolutions are provided. Then, the pre-processing steps needed for image fusion are stated. The commonly used protocols used for fusion are also described in this chapter. Furthermore, quantitative measures to assess fusion outcomes are mentioned.

CHAPTER 3

This chapter covers the three widely used methods of pansharpening or fusion techniques: Component Substitution, Multi-Resolution Analysis, and Model-Based methods. The general framework for each fusion method is discussed as well as its limitations and benefits. At the end of the chapter, an objective and a visual comparison are reported for representative fusion methods.

CHAPTER 4

In this chapter, the deep learning models which have been used for image fusion in remote sensing are stated. Then, the recently developed deep learning-based methods are discussed. Both single-objective and multi-objective loss functions are presented. Representative experimental results are reported in this chapter comparing deep learning-based approaches to conventional approaches.

CHAPTER 5

In this chapter, the unsupervised learning process for multispectral image fusion is covered. More specifically, the two major challenges in the training of deep learning models are discussed. These challenges include the availability of pseudo-ground truth data and dependency on the conventional loss function.

CHAPTER 6

Extensive experimental studies are reported in this chapter. The objective and subjective assessments are performed on different remote sensing datasets.

CHAPTER 7

This final chapter provides some anticipated future trends of image fusion in remote sensing.

1.3   REFERENCES

[1] Azarang, A. and Kehtarnavaz, N. 2020. Image fusion in remote sensing by multi-objective deep learning. *International Journal of Remote Sensing*, 41(24):9507–9524. DOI: 10.1080/01431161.2020.1800126. 1

[2] Sica, F., Pulella, A., Nannini, M., Pinheiro, M., and Rizzoli, P. 2019. Repeat-pass SAR interferometry for land cover classification: A methodology using Sentinel-1 Short-Time-Series. *Remote Sensing of Environment*, 232:111277. DOI: 10.1016/j.rse.2019.111277. 1

[3] Borsoi, R. A., Imbiriba, T., and Bermudez, J. C. M. 2019. Deep generative endmember modeling: An application to unsupervised spectral unmixing. *IEEE Transactions on Computational Imaging*, 6:374–384. DOI: 10.1109/tci.2019.2948726. 1

[4] Liu, S., Marinelli, D., Bruzzone, L., and Bovolo, F. 2019. A review of change detection in multitemporal hyperspectral images: Current techniques, applications, and challenges. *IEEE Geoscience and Remote Sensing Magazine*, 7(2):140–158. DOI: 10.1109/mgrs.2019.2898520. 1

[5] Chen, Y., Song, B., Du, X., and Guizani, M. 2019. Infrared small target detection through multiple feature analysis based on visual saliency. *IEEE Access*, 7:38996–39004. DOI: 10.1109/access.2019.2906076. 1

CHAPTER 2

Introduction to Remote Sensing

2.1 BASIC CONCEPTS

2.1.1 SPATIAL RESOLUTION

Spatial resolution in remote sensing denotes how much spatial detail in an image is visible. The capability to see small details is governed by spatial resolution. In other words, spatial resolution reflects the smallest object that may be resolved through the sensor, or the smallest ground region imaged with the aid of the instantaneous field of view (IFOV) of the sensor, or the smallest measurement on the ground defined by a pixel. What is meant by higher spatial resolution of an image in comparison to another image is that the first image contains more pixels than the second image for the same physical size. The quality of an image is determined by its spatial resolution or how much of the details of objects can be seen. For example, a spatial resolution of 10 m means that one pixel represents an area 10×10 m on the ground. Figure 2.1 shows an illustration of different spatial resolution in the QuickBird remote sensing dataset [1]. This dataset was captured in 2001 from the Sundarbans region in Bangladesh.

2.1.2 SPECTRAL RESOLUTION

Spectral resolution refers to the sensor ability to collect information at a specific wavelength of the spectrum. A finer spectral resolution means a narrower wavelength range for a specific channel or band. Spatial and spectral resolutions are closely connected to each other, i.e., a coarser spectral resolution automatically results in better spatial resolution and vice versa [2]. A higher spectral resolution leads to a better ability to see differences in the spectral content. For instance, a PAN image is indicative of a wide range of wavelengths. An object that exhibits high energy in the green portion of the visible band becomes indistinguishable in a PAN image from an object that exhibits the same amount of energy in the red band. The width, number, and position of spectral bands indicate the degree at which individual objects can be discriminated. The discrimination power of a multispectral image is higher than any single band image. Figure 2.2 provides a visual depiction of spectral resolution.

Spectral signatures or spectral reflectance curves of different ground objects can provide useful information, for example, the mineral content of rocks, the proper time for planting, the composition of buildings, soil moisture, and many other useful information.

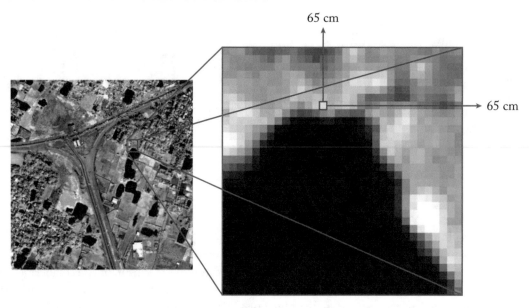

Figure 2.1: Spatial resolution (QuickBird dataset).

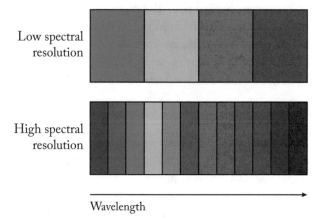

Figure 2.2: Low and high spectral resolutions.

2.1.3 RADIOMETRIC RESOLUTION

Radiometric resolution corresponds to the sensor ability to distinguish more gray-level or intensity values [3]. A sensor with a higher radiometric resolution captures more intensity levels. The maximum number of gray-level values available depends on the number of bits utilized in image capture and storage. The intrinsic radiometric resolution in remote sensing highly depends on the signal to ratio of the detection sensor. Radiometric resolution is defined by the number

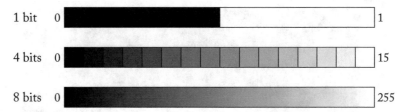

Figure 2.3: Radiometric resolution indicating number of gray-level or intensity values.

Table 2.1: Revisit time for earth observation satellites

Sensor	Revisit Period (in Days)
QuickBird	1–3.5
Landsat	16
Aster	16
IKONOS	3
GeoEye-1 and 2	2–8
WorldView-1 and 2	1.7
SPOT-5	2–3
SPOT-6 and 7	Daily
FORMOSAT-2	Daily
Pleiades-1A and 1B	Daily
Rapid Eye	Daily

of quantization levels utilized to digitize continuous intensity values. Thus, if a sensor utilizes 11 bits for data capture, there would be $2^{11} = 2048$ gray-level values ranging from 0–2047. Figure 2.3 provides a depiction of radiometric resolution represented by different gray-level or intensity values.

2.1.4   TEMPORAL RESOLUTION

Temporal resolution of a sensor may vary from hours to days. It is defined as the amount of time for a sensor to revisit the exact location on the ground [2, 3]. This time depends on the orbital properties as well as the sensor characteristics. For example, a sun-synchronous orbit follows sun illumination and thus the capturing of image data of a specific location remains the same every day. This is an important feature for visible-infrared sensors since it maximizes the temporal resolution of the sensor. Table 2.1 shows different temporal resolutions of a number of satellites [4].

MS   PAN

Figure 2.4: Registered MS and PAN images.

2.2 PRE-PROCESSING STEPS FOR IMAGE FUSION

2.2.1 IMAGE REGISTRATION

A critical pre-processing step before carrying out image fusion is image registration. The goal in image registration is to geometrically align one image to another and it is a crucial step for all image fusion approaches [3]. Considering that data are captured from the ground surface via different sensors, it is required to register them. An important issue to consider in image registration is finding the best geometric transformation for alignment, i.e., determining the best mapping function from one specific region in one image into its corresponding region in the reference image. Image registration methods used in remote sensing can be classified into two types [2]: area-based registration and feature-based registration. In area-based registration, the aim is to ensure that different sensors are pointing to the same region. Feature-based registration involves mapping different objects and aligning them together. An example of registered MS and PAN images is shown in Figure 2.4.

2.2.2 HISTOGRAM MATCHING

Histogram matching is used to put the digital values of two different images in the same range. From a statistical perspective, histogram matching results in matching the means and standard deviations of two images. It has been shown that histogram matching has a significant impact on the fusion outcome [2, 3]. In fact, the spectral distortion of a fused image is decreased by applying histogram matching. For example, it has been shown that in the Intensity-Hue-Saturation (IHS) representation, if the intensity component is replaced by the histogram matched PAN image rather than the original PAN image the fused image will have less spectral distortion. An example of histogram matching is shown in Figure 2.5. The following equation is used to match

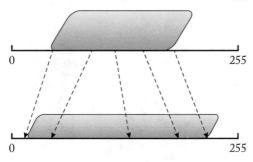

Figure 2.5: Histogram matching.

the histogram of a PAN image to the intensity component (I) of an IHS image:

$$P = \frac{\sigma_I}{\sigma_P} (P - \mu_P) + \mu_I,$$ (2.1)

where σ and μ represent the standard deviation and mean values, respectively.

2.3   FUSION PROTOCOLS

Due to the unavailability of high-resolution MS images to assess fusion outcomes, protocols have been established in the remote sensing literature for this purpose. These protocols are used to assess the integrity of the spectral and spatial information in the fused image. In what follows, the two most widely used protocols are stated.

2.3.1   REDUCED-RESOLUTION PROTOCOL (WALD'S PROTOCOL)

The protocol operating at the reduced resolution is based on the Wald's protocol in which the following conditions are considered [5].

1. The fused (pansharpened) image needs to be as close or similar as possible to the original MS image if resampled to a lower MS resolution.

2. The fused (pansharpened) image is desired to be as close or comparable as feasible to the original MS image with the highest spatial resolution.

3. The fused image set ought to be as close or comparable as possible to the intact MS image set with the highest spatial resolution.

With the assumption that the performance of fusion methods remain unchanged across different scales, both original PAN and MS images can be degraded to a coarser resolution. The original MS image is kept intact and the fusion process is performed in the down-scaled version of input images. This way, full-reference objective metrics are used for assessment. The flowchart of the Wald's protocol appears in Figure 2.6.

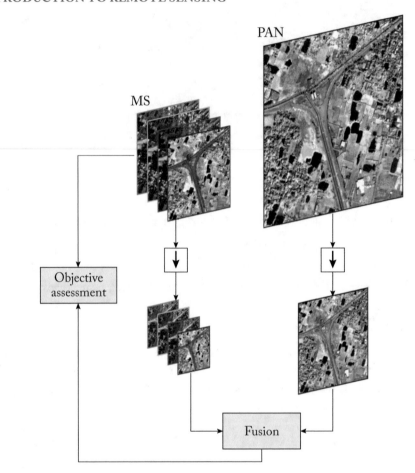

Figure 2.6: Wald's protocol.

2.3.2 FULL-RESOLUTION PROTOCOL (ZHOU'S PROTOCOL)

This protocol is performed at the full resolution mode of input images and separates spectral and spatial contents [6]. For spectral content, the assumption is based on the similarity between the spectral content of the fused image and the original MS image. In fact, based on this protocol, the spectral signature of different targets are kept the same. To compute spectral distortion, the average differences between pixel values of the fused image and the resampled MS image is computed band by band as follows:

$$\mathbf{D}_i = \frac{1}{m \times n} \sum \sum \left(\left| \widehat{\mathbf{F}}_i - \widehat{\mathbf{M}}_i \right| \right), \tag{2.2}$$

where m and n represent the size of images (number of rows and columns), and $\mathbf{F}_i$ and $\mathbf{M}_i$ denote the fused image and the resampled MS image, respectively. In order to assess the spatial content in this protocol, the assumption made is that the high frequency information in the PAN image is unique. Then, the high frequency of the fused image is compared. In fact, a high-pass Laplacian filter is applied to the PAN and the fused images, and then the correlation between these two images is used to serve as a spatial distortion measure.

2.4   QUANTITY ASSESSMENTS OF FUSION OUTCOMES

Naturally, better modeling of the human vision system (HVS) enables the development of more effective metrics. The following subsections cover the most widely used objective metrics that are used for the assessment of fused images or outcomes [7].

2.4.1   REDUCED-RESOLUTION METRICS

In this section, the reduced-resolution metric for the objective assessment of fusion outcomes is stated. Here, the original MS image is considered to be the reference image for spectral analysis while the PAN image is considered for spatial analysis. The most widely used objective metrics for reduced-resolution are as follows.

Spectral Angle Mapper (SAM):   This metric is based on the assumption that a single pixel in remote sensing images represents one specific ground cover index, and is thus devoted to only one particular class [8]. The color differences between the fused and MS images are characterized by this metric; see Figure 2.7. SAM has local and global values. The local value is computed as a map using the angle difference between each pixel of the fused image and its corresponding pixel in the MS image. Then, the difference map values are linearized between 0 and 255. The following equation is used to compute the local SAM metric:

$$\text{SAM}(x, y) = \frac{\langle F, M \rangle}{\|F\|_2 \|M\|_2}, \tag{2.3}$$

where F and M are the pixels of the fused image and the original MS image, respectively. On the other hand, the global value of SAM is computed by taking an average of all the pixels in the SAM map. It is represented in degree (°) or radian. The optimal value for global SAM is zero which means no color distortion in the fused image. One of the main drawbacks of the SAM metric is its inability to distinguish between positive and negative correlations since only absolute values are computed.

Correlation Coefficient (CC):   This metric reflects the cross correlation between the fused and reference images [9]. The range for CC is $[-1, 1]$, where 1 means the highest correlation

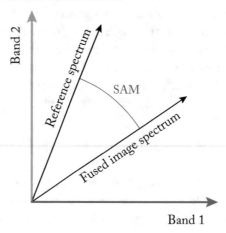

Figure 2.7: **SAM** concept.

between images. This metric is computed as follows:

$$CC = \frac{\sum_{x=1}^{N} \sum_{y=1}^{N} (\mathbf{F}(x, y) - \mu_{\mathbf{F}}) (\mathbf{M}(x, y) - \mu_{\mathbf{M}})}{\sqrt{\sum_{x=1}^{N} \sum_{y=1}^{N} (\mathbf{F}(x, y) - \mu_{\mathbf{F}})^2 (\mathbf{M}(x, y) - \mu_{\mathbf{M}})^2}}. \qquad (2.4)$$

This metric is first computed in a band-wise manner; then an average across bands is computed.

Universal Image Quality Index (UIQI): This is a similarity index which characterizes spectral and spatial distortions, and it is computed using the following equation [10]:

$$UIQI = \frac{\sigma_{\mathbf{F},\mathbf{M}}}{\sigma_{\mathbf{F}}\sigma_{\mathbf{M}}} \frac{2\mu_{\mathbf{F}}\mu_{\mathbf{M}}}{\mu_{\mathbf{F}}^2 + \mu_{\mathbf{M}}^2} \frac{2\sigma_{\mathbf{F}}\sigma_{\mathbf{M}}}{\sigma_{\mathbf{F}}^2 + \sigma_{\mathbf{M}}^2}. \qquad (2.5)$$

Equation (2.5) incorporates three distortions. The first term indicates the correlation loss, the second term the luminance loss, and the third term the contrast distortion [10]. The range for UIQI is [−1 1]. This metric has been among the most widely used metrics for assessing fusion outcomes. It provides a measure of structural distortion, thus it is more informative than the energy-based MSE metric.

Root Mean Square Error (RMSE): This widely used metric measures the spatial and spectral distortions in the fused outcome according to the following equation:

$$RMSE = \sqrt{\frac{\sum_{x=1}^{m} \sum_{y=1}^{n} (\mathbf{F}(x, y) - \mathbf{M}(x, y))}{m * n}}. \qquad (2.6)$$

The optimal value for RMSE is zero when the images **F** and **M** become identical.

Root Average Spectral Error (RASE):   Another global distortion metric is RASE. Low RASE values indicate more similarity between the fused outcome and the reference image. This measure incorporates mean values of individual bands according to the following equation:

$$\text{RASE} = \frac{100}{\sum_{i=1}^{N} \mu(i)} \sqrt{\frac{1}{N} \text{RMSE}(\mathbf{F}, \mathbf{M})}, \tag{2.7}$$

where $\mu(i)$ represents the mean of the i-th band. This measure is often stated as a percentage with the ideal value being zero.

Erreur Relative Globale Adimensionnelle de Synthése (ERGAS):   This metric denotes an improvement of the RASE metric by taking into consideration the scale ratio of the PAN and MS images. It reflects the global distortion in the fused outcome according to the following equation [11]:

$$\text{ERGAS} = 100 \frac{d_h}{d_l} \sqrt{\frac{1}{N} \sum_{i=1}^{N} \frac{\text{RMSE}(\mathbf{F}, \mathbf{M})}{\mu(i)}}, \tag{2.8}$$

where $\frac{d_h}{d_l}$ denotes the ratio between pixel sizes of PAN and MS images, e.g., $\frac{1}{4}$ for GeoEye-1 and QuickBird sensors. Low ERGAS values indicate less distortion in the fused outcome and more similarity in multispectral images.

Structure Similarity Index Metric (SSIM):   This metric provides the structural similarity between fused and MS images based on luminance, contrast, and structural cues according to the following equation [12]:

$$\text{SSIM} = \frac{(2\mu_{\mathbf{F}}\mu_{\mathbf{M}} + \epsilon_1)(2\sigma_{\mathbf{F}}\sigma_{\mathbf{M}} + \epsilon_1)}{(\mu_{\mathbf{F}}^2 + \mu_{\mathbf{M}}^2 + \epsilon_1)(\sigma_{\mathbf{F}}^2 + \sigma_{\mathbf{M}}^2 + \epsilon_2)}, \tag{2.9}$$

where ϵ_1 and ϵ_2 are small values to avoid a non-zero denominator. High SSIM values indicate more structural similarity between fused and MS images. This metric was derived by noting that structural similarities play a key role in the HVS.

2.4.2   FULL-RESOLUTION METRICS

Since in practice high-resolution MS images are not available and the fused outcome at the full-resolution needs to be assessed, full-resolution metrics are utilized. Next, a brief description of these metrics is stated.

Spectral Distortion Metric (D_λ):   This metric is computed between the low-resolution MS image and the fused outcome at the PAN image scale [7]. The UIQI metric between the MS bands, e.g., $\mathbf{M}_1$ and $\mathbf{M}_2$, is first computed and then subtracted from the corresponding multiplication at high-resolution (fused outcome, i.e., $\mathbf{F}_1$ and $\mathbf{F}_2$). This metric is expressed by the following equation:

$$D_\lambda = \sqrt[p]{\frac{1}{N(N-1)}\sum_{i=1}^{N}\sum_{j=1,j\neq i}^{N}\left|\text{UIQI}\left(\mathbf{M}_i,\mathbf{M}_j\right)-\text{UIQI}\left(\mathbf{F}_i,\mathbf{F}_j\right)\right|^p}. \qquad (2.10)$$

The exponent p is set to one by default but can be chosen to show larger differences between the two terms. Low D_λ metric values indicate less spectral distortion and the ideal value is zero.

Spatial Distortion Metric (D_s):　This metric consists of two terms. The first term is computed at low resolution between the UIQI of the original MS image and the degraded PAN image at the MS resolution. The second term is computed at the PAN image resolution between the UIQI of the fused image and the original PAN image. This metric is expressed by the following equation [7]:

$$D_s = \sqrt[q]{\frac{1}{N}\sum_{i=1}^{N}\left|\text{UIQI}\left(\mathbf{M}_i,\mathbf{P}^L\right)-\text{UIQI}\left(\mathbf{F}_i,\mathbf{P}\right)\right|^q}. \qquad (2.11)$$

The exponent q is set to one by default. The ideal value for D_s is zero which denotes no spatial distortion.

Quality of No Reference (QNR):　This metric involves a combination of the above two metrics. QNR is expressed by the following equation with the ideal value being one [7]:

$$\text{QNR} = (1-D_\lambda)^\alpha (1 - D_s)^\beta. \qquad (2.12)$$

The exponents α and β are control parameters to place the emphasis either on spectral or spatial distortion of the fused outcome. Normally, they are set to one.

2.5　REFERENCES

[1] Lasaponara, R. and Masini, N. 2007. Detection of archaeological crop marks by using satellite QuickBird multispectral imagery. *Journal of Archaeological Science*, 34(2):214–221. DOI: 10.1016/j.jas.2006.04.014. 3

[2] Pohl, C. and Van Genderen, J. 2016. *Remote Sensing Image Fusion: A Practical Guide*, CRC Press. DOI: 10.1201/9781315370101. 3, 5, 6

[3] Alparone, L., Aiazzi, B., Baronti, S., and Garzelli, A. 2015. *Remote Sensing Image Fusion*, CRC Press. DOI: 10.1201/b18189. 4, 5, 6

[4] https://www.satimagingcorp.com/satellite-sensors/ 5

[5] Wald, L., Ranchin, T., and Mangolini, M. 1997. Fusion of satellite images of different spatial resolutions: Assessing the quality of resulting images. *Photogrammetric Engineering and Remote Sensing*, 63(6):691–699. 7

[6] Zhou, J., Civco, D. L., and Silander, J. A. 1998. A wavelet transform method to merge Landsat TM and SPOT panchromatic data. *International Journal of Remote Sensing*, 19(4):743–757. DOI: 10.1080/014311698215973. 8

[7] Alparone, L. et al. 2008. Multispectral and panchromatic data fusion assessment without reference. *Photogrammetric Engineering and Remote Sensing*, 74(2):193–200. DOI: 10.14358/pers.74.2.193. 9, 11, 12

[8] Yuhas, R. H., Goetz, A. F. H., and Boardman, J. W. 1992. Discrimination among semi-arid landscape endmembers using the Spectral Angle Mapper (SAM) algorithm. *Proc. Summaries of the 3rd Annual JPL Airborne Geoscience Workshop*, pages 147–149. 9

[9] Azarang, A., Manoochehri, H. E., and Kehtarnavaz, N. 2019. Convolutional autoencoder-based multispectral image fusion. *IEEE Access*, 7:35673–35683. DOI: 10.1109/access.2019.2905511. 9

[10] Wang, Z. and Bovik, A. C. 2002. A universal image quality index. *IEEE Signal Processing Letters*, 9(3):81–84. DOI: 10.1109/97.995823. 10

[11] Wald, L. 2002. *Data Fusion: Definitions and Architectures—Fusion of Images of Different Spatial Resolutions*, Paris, France, Les Presses de lÉcole des Mines. 11

[12] Wang, Z., Bovik, A. C., Sheikh, H. R., and Simoncelli, E. P. 2004. Image quality assessment: From error visibility to structural similarity. *IEEE Transactions on Image Processing*, 13(4):600–612. DOI: 10.1109/tip.2003.819861. 11

CHAPTER 3

Conventional Image Fusion Approaches in Remote Sensing

3.1 COMPONENT SUBSTITUTION ALGORITHMS

This family of pansharpening or fusion algorithms uses the projection of a MS image into some other domain, with the idea that this transformation separates the spatial data from the spectral contents in specific spectral components [1]. The converted MS image is improved by means of substitution of the component containing the spatial data with the PAN image. It needs to be noted that the greater the correlation between the PAN image and the replaced component, the lower the distortion in the fused outcome [2]. A histogram matching of the PAN data to the replaced component is carried out before the replacement. As a result, the new PAN image exhibits the same mean and variance as the component substitution. The pansharpening process is performed by mapping the data to the original domain via an inverse transformation.

The above process is global (i.e., it performs on the entire image) with advantages and drawbacks. Component substitution (CS) approaches are generally categorized by their fidelity in the spatial details in the fused outcome [1], and they are, in general, fast or computationally efficient. On the other hand, they are no longer able to account for local dissimilarities among the PAN and MS images originated due to the spectral mismatch between the PAN and MS channels of the sensors, which might also produce substantial spectral distortions [3]. An alternative name of CS approaches is *projection substitution* [1] indicating the two main processing steps that are involved.

A new general equation of CS approaches was presented by Tu et al. [4] and then explored in later studies, e.g., [5, 6]. It is revealed that under the assumption of a linear transformation and the replacement of a single component, fusion can be achieved without the explicit computation of the forward and backward transformations, but via an appropriate injection framework. A general explanation of the CS fusion is given by:

$$\widehat{\mathbf{M}}_k = \widetilde{\mathbf{M}}_k + g_k(\mathbf{P} - \mathbf{I}_L) \tag{3.1}$$

in which the subscript k denotes the k-th spectral band, $\mathbf{g} = [g_1, \ldots, g_k, \ldots, g_N]$ is a vector of *injection gains*, while $\mathbf{I}_L$ is defined as

$$\mathbf{I}_L = \sum_{i=1}^{N} \omega_i \widetilde{\mathbf{M}}_k \tag{3.2}$$

in which the weight vector $\boldsymbol{\omega} = [\omega_1, \ldots, \omega_k, \ldots, \omega_N]$ is the first row of the transformation matrix and is selected to measure the degree of overlap between the PAN and MS data [7].

Figure 3.1 shows a flowchart denoting the fusion steps in CS approaches. More specifically, these steps perform: (1) interpolating the MS image for equivalent PAN scale; (2) computing the intensity term by Eq. (3.2); (3) histogram matching of intensity term and PAN image; and (4) injection of spatial details based on Eq. (3.1).

For CS approaches or methods, up/down scaling must guarantee the overlap of MS and PAN images at the finer resolution. Due to the acquisition geometry of the imaging sensors, interpolation with traditional zero-phase linear finite-impulse response filters might need realignment, e.g., via a bicubic filtering. Instead, linear nonzero-phase filters, having even numbers of coefficients, may be utilized [1].

The CS family also contains pansharpening approaches, called IHS, PCA, and GS methods [1], that vary by the mappings of the MS data utilized in the fusion process. As a result of the nonexistence of an exclusive function for extraction of the spatial component for replacement, several methods based on adaptive estimation have been introduced which are identified as *adaptive* CS [6]. In what follows, a more detailed description of representative CS methods is stated.

Generalized IHS (GIHS):   This method [4] extracts the transformation into the IHS color space that follows the HVS for processing intensity (I), hue (H), and saturation (S) contents. IHS images can be found from RGB images via a linear transformation, resulting in a drawback for processing MS images.

In [4], the IHS concept is generalized to images with more than three bands (GIHS). Later works [8] have demonstrated that GIHS can be formulated for any random set of non-negative weights as follows:

$$\widehat{\mathbf{M}}_k = \widetilde{\mathbf{M}}_k + \left(\sum_{i=1}^{N} \omega_i\right)^{-1} (\mathbf{P} - \mathbf{I}_L) \tag{3.3}$$

in which $\mathbf{I}_L$ follows from Eq. (3.2). Generally, the coefficients $\{\omega_i\}, k = 1, \ldots, N$, are equal to $1/N$. On the other hand, they can be adjusted based on the responses of the spectral channels [5]. The weight for spectral bands must be greater than zero and may not sum to one. The injection gains arrange for appropriate rescaling [8]. IHS ($N = 3$) might be employed as *fast* IHS, which evades a progressive calculation of the direct mapping, replacement, and the final backward stage. GIHS is fast since the transformation does not exist for $N > 3$.

In Eq. (3.1), considering the injection gains g_k, as follows:

$$g_k = \frac{\widetilde{\mathbf{M}}_k}{\mathbf{I}_L} \tag{3.4}$$

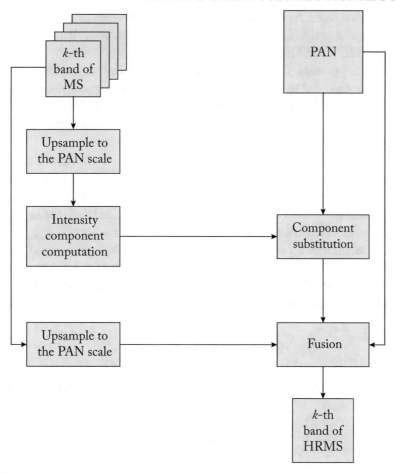

Figure 3.1: General CS framework.

results in

$$\widehat{\mathbf{M}}_k = \widetilde{\mathbf{M}}_k + \frac{\widetilde{\mathbf{M}}_k}{\mathbf{I}_L}\,(\mathbf{P} - \mathbf{I}_L) = \mathbf{M}_k \cdot \frac{\mathbf{P}}{\mathbf{I}_L} \tag{3.5}$$

which is known as the *Brovey transform* (BT) fusion algorithm [9]. BT might be acquired by the general model in Eq. (3.1) with the selection of injection gains in Eq. (3.4) which denotes a *multiplicative* fusion or a spatial modulation of spectral pixels.

The specific category of algorithms stated by Eq. (3.5), varying in the selection of spectral weights in Eq. (3.2), is called *relative spectral contribution* (RSC) [10]. However, as noted in [11], all the CS methods are considered in one class. Based on Eq. (3.1), RSC can be considered to be a specific case of the CS class since such methods can be exhibited the same features as CS methods [12–14].

Adaptive IHS (AIHS):   As mentioned earlier, the IHS fusion technique converts a color image from RGB space to the IHS color space. Then, the **I** (intensity) band is replaced by the PAN image. The intensity band **I** is obtained using Eq. (3.2).

In the fusion process, first the MS image is upsampled to the PAN size and each band of the image is normalized to the range [0, 1]. After completing the preprocessing steps, a histogram matching of the PAN image **P** is performed to assure that the μ and σ of the PAN image and the MS image are within the same range as per Eq. (2.1). Last, the fused product is made using Eq. (3.1).

The final fused IHS image usually has high spatial but low spectral resolution. The AIHS [14] is introduced to produce high spectral resolution by finding an image adaptive α coefficient. Also, it is possible to extract the edges of the PAN image and combine it with the MS image to increase spectral fidelity. At the final step, the AIHS involves these techniques to enhance the quality of the fused product.

In order to minimize spectral distortion in the IHS fused product, AIHS offers a modification on the IHS that differs in the manner by which the intensity band is computed depending on the initial MS and PAN images. To minimize spectral inconsistency, the intensity component needs to approximate the PAN data. In this AIHS method [14], the coefficients α is obtained as follows:

$$\mathbf{P} \approx \sum_{k=1}^{N} \alpha_k \widetilde{\mathbf{M}}_k. \tag{3.6}$$

In order to compute these coefficients, the following function G is minimized with respect to the coefficients:

$$\min_{\alpha} G\left(\alpha\right) = \sum_{k=1}^{N} \left(\alpha_k \widetilde{\mathbf{M}}_k - \mathbf{P}\right)^2 + \gamma \sum_{k=1}^{N} \left(\max\left(0, -\alpha_k\right)\right)^2. \tag{3.7}$$

Furthermore, the edges from the PAN image is transferred to the fused product using an edge detector. This approach extracts the edges from the PAN image. Where there are edges, the IHS method is imposed; otherwise, the MS image is used. The fused product $\widehat{\mathbf{M}}_k$ is formed as follows:

$$\widehat{\mathbf{M}}_k = \widetilde{\mathbf{M}}_k + h \odot \left(\mathbf{P} - \mathbf{I}\right) \tag{3.8}$$

in which h is an edge detecting function. This function is designed to become zero off edges and to 1 on edges. The edge extraction can be acquired using standard edge detection methods [15]. Empirically, it is seen that the best results are obtained by the edge detector developed by Perona and Malik [16]. In this detector, the edges of the PAN data are obtained by the following function:

$$h = \exp\left(-\frac{\lambda}{|\nabla\mathbf{P}|^4 + \epsilon}\right), \tag{3.9}$$

where $\nabla\mathbf{P}$ is the gradient of the PAN image, λ is a parameter controlling image smoothness and denotes how large the gradient needs to be in order to be an edge, and ε is a small value that ensures a nonzero denominator. The parameters are selected experimentally to be $\lambda = 10^{-9}$ and $\varepsilon = 10^{-10}$. Using these values and incorporating them into the edge detector with the original IHS significantly increases the spectral resolution.

Although the IHS approach exhibits high spatial efficiency, it is seriously affected by spectral distortion problems. Over the years, enhancement techniques have been built to address this disadvantage. For instance, the spectrally adjusted IHS (SAIHS) [1] approach is introduced on the basis of an observation of the reflectance spectra of the IKONOS sensor. SAIHS can only be made effective to some degree for IKONOS images, but not for other kinds of sensor images. There are also other IHS spectral response modification approaches that use the so-called vegetation index. The above-mentioned AIHS method has recently been formulated by changing the linear combination coefficients of the MS bands in the spatial information extraction step. However, in the absence of adaptability, the weights caused by the edges of the PAN image in the spatial information injection phase tend to be too high resulting in color distortion of the vegetation areas. In addition, it has been experimentally discovered that the weights caused by the edges of the MS bands will result in too smooth fused outcomes. Motivated by this observation and considering the need to keep the ratios between each pair of MS bands unchanged, the AIHS approach has been extended to incorporate an updated AIHS method known as IAIHS.

In order to move the edges from the PAN image to the fused product, the weighting matrix in the AIHS process is a function of the edges of the PAN image. Although the edges that appear in the PAN image cannot emerge in each MS band [1], it is not reasonable to inject the same amount of information into different bands simultaneously. This causes spectral distortion in the fused product. It is thus reasonable to expect that the weighting matrix that modulates each MS band should be different. Therefore, instead of using the edge detector on the PAN image singularly for each MS band, it has been proposed to use the following MS edge detector:

$$h_{\mathbf{M}_k} = \exp\left(-\frac{\lambda}{\left|\nabla \widetilde{\mathbf{M}}_k\right|^4 + \epsilon}\right). \tag{3.10}$$

It is seen that the AIHS method using the MS edge detector produces too smooth images which appear like an oil painting.

In order to assess the quality obtained by the PAN edge detector and the MS edge detector, it is fair to assume that the proper detection of the edge for each MS band, referred to as h_{total_k}, must be a trade-off among h and $h_{\mathbf{M}_k}$. In addition, in order to prevent spectral degradation, the proportions between each pair of MS bands must stay unchanged. Maintaining such proportions is of particular important for preserving the spectral details of the original MS image. To achieve this, the following condition has to be met:

$$h_{total_k} \propto \frac{\widetilde{\mathbf{M}}_k}{1/N \sum_{k=1}^{N} \widetilde{\mathbf{M}}_k}. \tag{3.11}$$

By taking into consideration the aforementioned assumptions, the following edge detector has been defined to control the amount of injected spatial details for each band:

$$h_{total_k} = \frac{\widetilde{\mathbf{M}}_k}{1/N \sum_{k=1}^{N} \widetilde{\mathbf{M}}_k} \left(\beta h + (1 - \beta) h_{\widetilde{\mathbf{M}}_k} \right) \tag{3.12}$$

in which β is a tradeoff parameter. Then, the final framework for the IAIHS method is stated as follows:

$$\widehat{\mathbf{M}}_k = \widetilde{\mathbf{M}}_k + h_{total_k} \odot (\mathbf{P} - \mathbf{I}). \tag{3.13}$$

Some improvement on the IAIHS method is seen in the recent literature. In fact, it has been found that the parameter β in Eq. (3.13) can be optimized by defining a proper single or multi-objective function. For example, in [17], the ERGAS metric is employed as an objective function to find the best values for β for each band separately. To solve the optimization problem, the particle swarm optimization (PSO) algorithm is used. As another example, in [18], the parameter β is optimized using a multi-objective function, where the ERGAS and CC metrics are considered together to minimize spectral distortion as well as spatial distortion.

Band-Dependent Spatial Detail (BDSD): The *band-dependent spatial detail* (BDSD) method [19], as shown in Figure 3.2, starts from an extended version of the generic formulation in Eq. (3.1) as follows:

$$\widehat{\mathbf{M}}_k = \widetilde{\mathbf{M}}_k + g_k \left(\mathbf{P} - \sum_{i=1}^{N} \omega_{k,i} \widetilde{\mathbf{M}}_k \right), \tag{3.14}$$

where the range of index k is from 1 to N. Equation (3.14) can be rewritten by incorporating the injection gains and the weights together as follows:

$$\lambda_{k,i} = \begin{cases} g_k & \text{if } i = N + 1 \\ -g_k \cdot \omega_{k,i} & \text{otherwise.} \end{cases} \tag{3.15}$$

Using this formula, Eq. (3.14) can be rewritten in a compact matrix form as follows:

$$\widehat{\mathbf{M}}_k = \widetilde{\mathbf{M}}_k + \mathbf{H}\lambda_k, \tag{3.16}$$

where $\mathbf{H} = \left[\widetilde{\mathbf{M}}_1, \ldots, \widetilde{\mathbf{M}}_N, \mathbf{P} \right]$, and $\lambda_k = \left[\lambda_{k,1}, \ldots, \lambda_{k,N+1} \right]^T$. It is worth mentioning that images here are turned into column vectors. The optimal MMSE joint estimation of the weights-and-gains vector λ would encompass the utilization of the unknown target image $\widehat{\mathbf{M}}$ and is therefore carried out at a reduced resolution mode. Consequently, the solution is found to be

$$\lambda_k = \left(\mathbf{H}_d^T \mathbf{H}_d \right)^{-1} \mathbf{H}_d^T \left(\widetilde{\mathbf{M}}_k - \widetilde{\mathbf{M}}_k^{LP} \right), \tag{3.17}$$

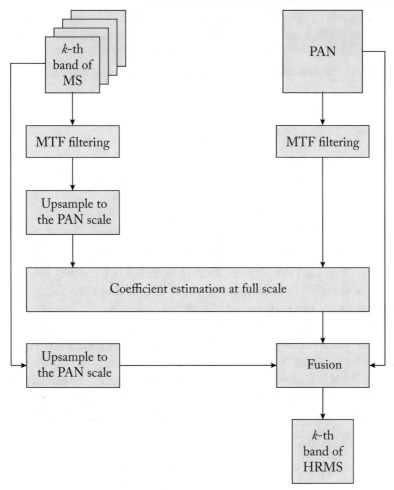

Figure 3.2: BDSD framework.

where $\mathbf{H}_d$ is the reduced scale of $\mathbf{H}$, $\widetilde{\mathbf{M}}_k^{LP}$ is the low pass version of $\widetilde{\mathbf{M}}_k$, in which the spatial frequency response and the *modulation transfer function* (MTF) of the MS sensor are matched. Several improvements have been injected into the BDSD framework. The generalized framework of the BDSD [1] is expressed as

$$\widehat{\mathbf{M}}_k = \widetilde{\mathbf{M}}_k + g_k \left(\mathbf{P} - \sum_i^N \omega_{k,i} \widetilde{\mathbf{M}}_k - \beta_k \widetilde{P} \right). \qquad (3.18)$$

In this equation, the term $\widetilde{P}$ denotes the LRPAN image. Equation (3.2) incorporates the Component Substitution (CS) and Multi-Resolution Analysis (MRA) frameworks in one place. Recently, a modification [20] has been applied to the BDSD framework in which the intensity components of each MS band is derived through a nonlinear estimation of joint terms. Specifically, the detail map of each MS band is taken from the other spectral bands and their combined multiplications. Mathematically, in this modification, the detail map is computed as follows:

$$\mathbf{D}_k = \mathbf{P} - \mathbf{I}_k \tag{3.19}$$

and

$$\mathbf{I}_k = \sum_{\substack{i=1 \\ i \neq k}}^{N} \omega_{i,k} \widetilde{\mathbf{M}}_i + \sum_{\substack{j=1 \\ j \neq k}}^{N} b_{j,k} \widetilde{\mathbf{M}}_j \odot \widetilde{\mathbf{M}}_k, \tag{3.20}$$

where $\mathbf{D}_k$ denotes the detail map of the k-th spectral band, $w_{i,k}$'s and $b_{j,k}$'s represent weights, and the notation $\odot$ means element-wise multiplication of two spectral band images. The range of the weights is set to $[-1, 1]$ in order to avoid distortion in the fusion outcomes. The last index of weights shows the k-th spectral band. Since the spectral contents of MS bands have some overlap, the last term in Eq. (3.20) takes into consideration the joint multiplications of LRMS bands to make the estimation of the detail map more accurate. As described in [14], the weights can be acquired by minimizing the Mean Square Error (MSE) between the MTF matched to the PAN image $\mathbf{P}$ and the intensity component $\mathbf{I}_k$ for each spectral band [1]. In fact, the low frequency components of the PAN image is estimated using the LRMS bands separately. Mathematically, the intensity component of each spectral band (expanded version of Eq. (3.20)) is derived via the following equations (in a 4-band system):

$$\begin{aligned}
\mathbf{I}_1 &= w_{2,1}\widetilde{\mathbf{M}}_2 + w_{3,1}\widetilde{\mathbf{M}}_3 + w_{4,1}\widetilde{\mathbf{M}}_4 + b_{2,1}\widetilde{\mathbf{M}}_2 \odot \widetilde{\mathbf{M}}_1 \\
&\quad + b_{3,1}\widetilde{\mathbf{M}}_3 \odot \widetilde{\mathbf{M}}_1 + b_{4,1}\widetilde{\mathbf{M}}_4 \odot \widetilde{\mathbf{M}}_1 \\
\mathbf{I}_2 &= w_{1,2}\widetilde{\mathbf{M}}_1 + w_{3,2}\widetilde{\mathbf{M}}_3 + w_{4,2}\widetilde{\mathbf{M}}_4 + b_{1,2}\widetilde{\mathbf{M}}_1 \odot \widetilde{\mathbf{M}}_2 \\
&\quad + b_{3,2}\widetilde{\mathbf{M}}_3 \odot \widetilde{\mathbf{M}}_2 + b_{4,2}\widetilde{\mathbf{M}}_4 \odot \widetilde{\mathbf{M}}_2 \\
\mathbf{I}_3 &= w_{1,3}\widetilde{\mathbf{M}}_1 + w_{2,3}\widetilde{\mathbf{M}}_2 + w_{4,3}\widetilde{\mathbf{M}}_4 + b_{1,3}\widetilde{\mathbf{M}}_1 \odot \widetilde{\mathbf{M}}_3 \\
&\quad + b_{2,3}\widetilde{\mathbf{M}}_2 \odot \widetilde{\mathbf{M}}_3 + b_{4,3}\widetilde{\mathbf{M}}_4 \odot \widetilde{\mathbf{M}}_3 \\
\mathbf{I}_4 &= w_{1,4}\widetilde{\mathbf{M}}_1 + w_{2,4}\widetilde{\mathbf{M}}_2 + w_{3,4}\widetilde{\mathbf{M}}_3 + b_{1,4}\widetilde{\mathbf{M}}_1 \odot \widetilde{\mathbf{M}}_4 \\
&\quad + b_{2,4}\widetilde{\mathbf{M}}_2 \odot \widetilde{\mathbf{M}}_4 + b_{3,4}\widetilde{\mathbf{M}}_3 \odot \widetilde{\mathbf{M}}_4.
\end{aligned} \tag{3.21}$$

The flowchart of this approach is depicted in Figure 3.3. As normally done, after obtaining the detail map of each LRMS band, the fusion is achieved via the following equation:

$$\widehat{\mathbf{M}}_k = \widetilde{\mathbf{M}}_k + g_k \mathbf{D}_k. \tag{3.22}$$

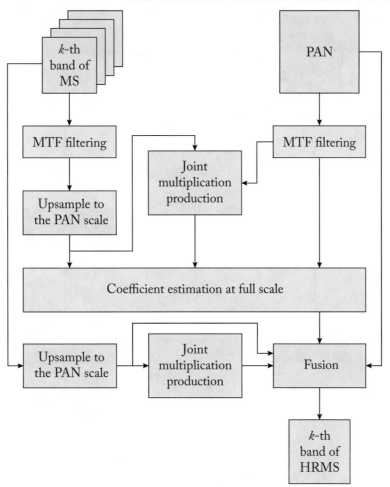

Figure 3.3: The nonlinear BDSD framework.

Here, it is worth mentioning that the injection gains g_k are computed using the following equation as discussed in [1]:

$$g_k = \frac{cov\left(\widetilde{\mathbf{MS}}_k, \mathbf{I}_k\right)}{var(\mathbf{I}_k)}. \tag{3.23}$$

3.2   MULTI-RESOLUTION ANALYSIS ALGORITHMS

In the second class of pansharpening methods, the contribution of the PAN image to the fused product is made by computing the difference between $\mathbf{P}$ and a low-pass version $\mathbf{P}_L$; see Fig-

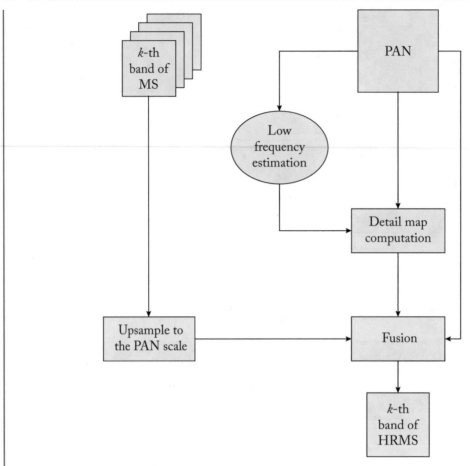

Figure 3.4: General MRA framework.

ure 3.4. Namely, the HRMS image is defined as

$$\widehat{\mathbf{M}}_k = \widetilde{\mathbf{M}}_k + g_k(\mathbf{P} - \mathbf{P}_L). \tag{3.24}$$

In [21], this paradigm has been denoted as *Amélioration de la Résolution Spatiale par In-jection de Structures* (ARSIS). The objectives of these methods are the preservation of the entire information of the LRMS image and the addition of features taken from the PAN data via filtering in spatial domain.

Based on Tu et al. [4], different approaches belonging to this category of pansharp-ening methods are classified by the way of acquiring the image $\mathbf{P}_L$ and the injection gains $\{g_k\}, k = 1, \ldots, N$. As a general framework, $\mathbf{P}_L$ is obtained through a recursive decomposition (called MRA) that targets constructing a series of images with increasingly reduced information through the repeated application of some operators. Information is observed as spatial frequency, and therefore, possibly increased sampling step sizes may be used through the sequence. That is why sometimes the name *pyramid* is used.

The type of decomposition ranges from simple methods based on a single level decomposition acquired through a simple LPF to more complex methods relying upon a formal MRA. The conventional methods in MRA is briefly described next.

High-Pass Filtering (HPF):   One of the early studies to find $\mathbf{P}_L$ is to apply a low-pass filter to extract the low-frequency information of the PAN image. Then, the high frequency information which relates to the missing detail of LRMS is obtained by subtracting the low-pass filtered PAN from its original image. Mathematically, if h_{LP} is considered as a low-pass filter, Eq. (3.24) can be written as follows:

$$\widehat{\mathbf{M}}_k = \widetilde{\mathbf{M}}_k + g_k(\mathbf{P} - h_{LP} * \mathbf{P}) \tag{3.25}$$

where $*$ denotes the convolution operation. This process can be performed by several filters, e.g., Laplacian and Gaussian.

Smoothing Filter-based Intensity Modulation (SFIM):   Smoothing Filter-based Intensity Modulation (SFIM) is an approach which is based on the idea of using the ratio between HRPAN and LRPAN obtained by low-pass filtering. Without changing the spectral content of LRMS, the spatial information of the PAN image is injected into the LRMS to obtain HRMS using the following equation:

$$\widehat{\mathbf{M}}_k = \widetilde{\mathbf{M}}_k + \frac{\widetilde{\mathbf{M}}_k}{\mathbf{P}_L}(\mathbf{P} - \mathbf{P}_L) = \frac{\widetilde{\mathbf{M}}_k}{\mathbf{P}_L} \cdot \mathbf{P}. \tag{3.26}$$

Indusion: A multispectral image fusion method derived from the so-called induction scaling technique. The induction method deliberates the up-scaling (enlargement) as the inverse problem of down-scaling (reduction). This results in a new image, when reduced it provides a primary image. This process is called *reduction constraint*. If $\mathbf{I}$ denotes the primary image, R the reduction filter, $\mathbf{I}^{\frac{1}{\alpha}}$ the upsampled image, and α the reduction ratio, then the reduction limit can be formulated as $\left[\mathbf{I}^{\frac{1}{\alpha}} * R\right] \downarrow \alpha = \mathbf{I}$. However, for a given image $\mathbf{I}$ and a reduction filter R, there is a set of enlarged images that meets the reduction constraint. This set of images ($\Omega_I = \{\mathbf{X} | [\mathbf{X} * R] \downarrow \alpha = \mathbf{I}\}$) is named induction set. The induction set needs to be searched. Induction simply involves a mapping J onto Ω so as to obtain an induced image K that belongs to Ω. In [22], a computationally efficient non-recursive implementation of induction is outlined as follows:

$$\mathbf{K} = \mathbf{J} + [\mathbf{I} - [\mathbf{J} * R] \downarrow a] \uparrow a * A. \tag{3.27}$$

The Indusion fusion method can be performed based on induction and Eq. (3.27) can be rewritten as follows:

$$\mathbf{K} = \mathbf{J} - [[\mathbf{J} * R] \downarrow a] \uparrow a * A + [\mathbf{I}] \uparrow a * A. \tag{3.28}$$

The first term of Eq. (3.28) denotes the high frequency content, spatial edges, and the second part of the equation represents the low frequency content of the image. The idea behind the indusion in pansharpening is to substitute image **J** by the PAN image noting that the high frequency details of the PAN has to be added to the LRMS. The pansharpening version of Eq. (3.28) can be stated as follows:

$$K = \mathbf{P} - [[\mathbf{P} * R] \downarrow a] \uparrow a * A + [I] \uparrow a * A. \tag{3.29}$$

Pyramidal Decompositions: The reduction in resolution needed to obtain a low-pass $\mathbf{P}_L$ version on the original scale of MS data can be achieved in one or more steps, i.e., by implementing a specific LPF with a cut-frequency equal to $1/L$ and decimating by L or by several fractional blocks. The second approach, which involves the first as a special case, is generally referred to as pyramidal decomposition and goes all the way back to the pioneer efforts of Burt and Adelson [23], in which Gaussian LPFs are used to conduct analysis. The corresponding differential version, obtained by measuring the variations among the Gaussian pyramid levels, is called LP and was later shown to be very useful for multispectral image fusion. In particular, Gaussian filters can be fine-tuned to closely fit the MTF sensor. This allows the extraction of information from the PAN image that are not available to the MS sensor. Since the Gaussian filter or mask is defined by a single parameter, its frequency response is fully known. For this purpose, the value of the amplitude response at the Nyquist frequency is used, which is commonly reported by manufacturers as a sensor specification. At the same time, it is useful to keep in mind that component aging can have a significant impact on this parameter.

The methods in this category are MTF-GLP with HPM (MTF-GLP-HPM), MTF Generalized LP (MTF-GLP), and MTF-GLP-HPM method, followed by a postprocessing (PP) phase aimed at correcting the noise generated in the absence of continuities. The widely used injection gains for MRA can be obtained through the following equation depending on $\mathbf{P}_L$ for each spectral band:

$$g_k = \frac{cov\left(\widetilde{\mathbf{M}}_k, \mathbf{P}_L\right)}{var\left(\mathbf{P}_L\right)}. \tag{3.30}$$

This injection rule can also be applied in the GS algorithm to improve the fusion results, known as MTF-GLP with context based decision (MTF-GLP-CBD), since the injection coefficients can be locally optimized using patch-wise vectors over nonoverlapping areas.

The undecimated "à trous" approach [24] has recently gained much attention. In practice, even though the non-orthogonality (which indicates that the wavelength plane must retain details for the adjacent plane) could affect the spectral efficiency of the fused outcome [25], its practical characteristics, such as the shift-invariance property [26] and the capability to be easily calibrated to the MTF sensor [1], provide reliable pansharpened images. A commonly used implementation of the "à trous" filter in the vertical and horizontal directions of 1D filters is given by

$$h = [1\ 4\ 6\ 4\ 1]. \tag{3.31}$$

In this case, known as ATWT, choosing the signal-dependent injection formula in Eq. (3.26) for obtaining the final fused product results in the formulation indicated in [27]. Moreover, other implementations using *Model 2* and *Model 3* [21] are ATWT-M2 and ATWT-M3, respectively. They comprise reconciling the first two moments of the data derived from the PAN image. The mean and standard deviations of the MS and PAN wavelet coefficients at the original scale of MS are utilized in all these cases. Although Model 2 requires a deterministic correlation between the respective quantities, Model 3 is regularized for the least square fitting.

All the previous wavelet methods are based on the choice of unitary injection coefficients $\{g_k\}$, $k = 1, \ldots, N$. However, some further improvements can be achieved by injecting the details using the HPM paradigm in Eq. (3.26) [28]. As an example of wavelet-based methods employing a different choice, the *additive wavelet luminance proportional* (AWLP) was implemented in [29] by using the more general fusion formula reported in Eq. (3.26) with the injection coefficients defined as

$$g_k = \frac{\widetilde{\mathbf{M}}_k}{\sum_{k=1}^{N} \widetilde{\mathbf{M}}_k}. \tag{3.32}$$

3.3   MODEL-BASED ALGORITHMS

The majority of multispectral image fusion methods fall into the CS and MRA pansharpening frameworks. However, there are some methods that fall outside these frameworks. Among them, the model-based optimization methods are of prominence, which are mentioned in this section. These methods involve a regularization optimization problem by minimizing a loss function according to a model of the imaging sensor or of the relationship between the captured low- and high-resolution images.

This category of pansharpening methods use a model based on a prior knowledge of the ground scene. Usually, an observational model is assumed for noisy satellite images as follows [30]:

$$y_i = \mathbf{M}_i \mathbf{B}_i x_i + \epsilon_i, \quad i = 1, 2, \ldots, L \tag{3.33}$$

in which $\mathbf{M}_i$ denotes the downsampling operator, $\mathbf{B}_i$ a circulant blurring matrix, x_i the vecotorized target image, L the total number of bands, and ϵ_i the noise model. This model is in fact a generalized Bayesian model reflecting prior distribution of the captured data. To accurately define the imaging steps of satellite sensors, a MTF for each channel is needed. MTF-shaped kernels with different cut-off frequencies are utilized to implement the blurring operation as low-pass filters. Then, the pan-sharpening problem becomes obtaining an estimation of a HRMS image x_i via using y_i and $\mathbf{P}$. The degraded model plays a critical role in preserving spectral consistency in the fused outcome. It is of practical significance to note that the ill-posed inverse problem stated in Eq. (3.33) has no unique solution. To obtain an accurate and unique solution, fine-tuning terms that indicate structural prior knowledge of the target HRMS image are normally considered.

3.4 REFERENCES

[1] Vivone, G., Alparone, L., Chanussot, J., Dalla Mura, M., Garzelli, A., Licciardi, G. A., Restaino, R., and Wald, L. 2014. A critical comparison among pan-sharpening algorithms. *IEEE Transactions on Geoscience and Remote Sensing*, 53(5):2565–2586. DOI: 10.1109/tgrs.2014.2361734. 15, 16, 19, 21, 22, 23, 26

[2] Rahmani, S., Strait, M., Merkurjev, D., Moeller, M., and Wittman, T. 2010. An adaptive IHS pan-sharpening method. *IEEE Geoscience and Remote Sensing Letters*, 7(4):746–750. DOI: 10.1109/lgrs.2010.2046715. 15

[3] Aiazzi, B., Alparone, L., Baronti, S., Carlà, R., Garzelli, A., and Santurri, L. 2016. Sensitivity of pan-sharpening methods to temporal and instrumental changes between multispectral and panchromatic data sets. *IEEE Transactions on Geoscience and Remote Sensing*, 55(1):308–319. DOI: 10.1109/tgrs.2016.2606324. 15

[4] Tu, T.-M., Su, S.-C., Shyu, H.-C., and Huang, P. S. 2001. A new look at IHS like image fusion methods. *Information Fusion*, 2(3):177–186. DOI: 10.1016/s1566-2535(01)00036-7. 15, 16, 24

[5] Tu, T.-M., Huang, P. S., Hung, C.-L., and Chang, C.-P. 2004. A fast intensityhue-saturation fusion technique with spectral adjustment for IKONOS imagery. *IEEE Geoscience and Remote Sensing Letters*, 1(4):309–312. DOI: 10.1109/lgrs.2004.834804. 15, 16

[6] Aiazzi, B., Baronti, S., and Selva, M. 2007. Improving component substitution pan-sharpening through multivariate regression of MS+Pan data. *IEEE Transactions on Geoscience and Remote Sensing*, 45(10):3230–3239. DOI: 10.1109/tgrs.2007.901007. 15, 16

[7] Azarang, A. and Kehtarnavaz, N. 2020. Multispectral image fusion based on map estimation with improved detail. *Remote Sensing Letters*, 11(8):797–806. DOI: 10.1080/2150704x.2020.1773004. 16

[8] Dou, W., Chen, Y., Li, X., and Sui, D. 2007. A general framework for component substitution image fusion: An implementation using fast image fusion method. *Computers and Geoscience*, 33(2):219–228. DOI: 10.1016/j.cageo.2006.06.008. 16

[9] Gillespie, A. R., Kahle, A. B., and Walker, R. E. 1987. Color enhancement of highly correlated images—II. Channel ratio and Chromaticity, trans-form techniques. *Remote Sensing of Environment*, 22(3):343–365. DOI: 10.1016/0034-4257(87)90088-5. 17

[10] Amro, I., Mateos, J., Vega, M., Molina, R., and Katsaggelos, A. K. 2011. A survey of classical methods and new trends in pan-sharpening of multispectral images. *EURASIP Journal on Advances in Signal Processing*, 2011(1):79:1–79:22. DOI: 10.1186/1687-6180-2011-79. 17

[11] Aiazzi, B., Alparone, L., Baronti, S., Garzelli, A., and Selva, M. 2012. Twenty-five years of pan-sharpening: A critical review and new developments. *Signal and Image Processing for Remote Sensing*, 2nd ed., Chen, C.-H., Ed. Boca Raton, FL, CRC Press, pages 533–548. DOI: 10.1201/b11656-30. 17

[12] Baronti, S., Aiazzi, B., Selva, M., Garzelli, A., and Alparone, L. 2011. A theoretical analysis of the effects of aliasing and misregistration on pansharpened imagery. *IEEE Journal of Select Topics in Signal Processing*, 5(3):446–453. DOI: 10.1109/jstsp.2011.2104938. 17

[13] Laben, C. A. and Brower, B. V. 2000. Process for enhancing the spatial resolution of multispectral imagery using pan-sharpening. U.S. Patent6 011 875. 17

[14] Rahmani, S., Strait, M., Merkurjev, D., Moeller, M., and Wittman, T. 2010. An adaptive IHS pan-sharpening method. *IEEE Geoscience and Remote Sensing Letters*, 7(4):746–750. DOI: 10.1109/lgrs.2010.2046715. 17, 18, 22

[15] Canny, J. 1986. A computational approach to edge detection. *IEEE Transactions on Pattern Analysis and Machine Intelligence*, PAMI-8(6):679–698. DOI: 10.1109/tpami.1986.4767851. 18

[16] Perona, P. and Malik, J. 1990. Scale-space and edge detection using anisotropic diffusion. *IEEE Transactions on Pattern Analysis and Machine Intelligence*, 12(7):629–639. DOI: 10.1109/34.56205. 18

[17] Azarang, A. and Ghassemian, H. 2017. An adaptive multispectral image fusion using particle swarm optimization. *Iranian Conference on Electrical Engineering (ICEE)*, IEEE, pages 1708–1712. DOI: 10.1109/iraniancee.2017.7985325. 20

[18] Khademi, G. and Ghassemian, H. 2017. A multi-objective component-substitution-based pan-sharpening. *3rd International Conference on Pattern Recognition and Image Analysis (IPRIA)*, IEEE, pages 248–252. DOI: 10.1109/pria.2017.7983056. 20

[19] Garzelli, A., Nencini, F., and Capobianco, L. 2008. Optimal MMSE pan sharpening of very high resolution multispectral images. *IEEE Transactions on Geoscience Remote Sensing*, 46(1):228–236. DOI: 10.1109/tgrs.2007.907604. 20

[20] Azarang, A. and Kehtarnavaz, N. 2020. Multispectral image fusion based on map estimation with improved detail. *Remote Sensing Letters*, 11(8):797–806. DOI: 10.1080/2150704x.2020.1773004. 22

[21] Ranchin, T. and Wald, L. 2000. Fusion of high spatial and spectral resolution images: The ARSIS concept and its implementation. *Photogrammatic Engineering and Remote Sensing*, 66(1):49–61. 24, 27

[22] Khan, M. M., Chanussot, J., Condat, L., and Montavert, A. 2008. Indusion: Fusion of multispectral and panchromatic images using the induction scaling technique. *IEEE Geoscience and Remote Sensing Letters*, 5(1):98–102. DOI: 10.1109/lgrs.2007.909934. 25

[23] Burt, P. J. and Adelson, E. H. 1983. The Laplacian pyramid as a compact image code. *IEEE Transactions on Communications*, COM-31(4):532–540. DOI: 10.1109/tcom.1983.1095851. 26

[24] Shensa, M. J. 1992. The discrete wavelet transform: Wedding the à trousand Mallat algorithm. *IEEE Transactions on Signal Processing*, 40(10):2464–2482. DOI: 10.1109/78.157290. 26

[25] González-Audícana, M., Otazu, X., Fors, O., and Seco, A. 2005. Comparison between Mallat's and the "à trous" discrete wavelet transform based algorithms for the fusion of multispectral and panchromatic images. *International Journal of Remote Sensing*, 26(3):595–614. DOI: 10.1080/01431160512331314056. 26

[26] Vetterli, M. and Kovacevic, J. 1995. *Wavelets and Subband Coding*, Englewood Cliffs, NJ, Prentice Hall. 26

[27] Núñez, J. et al. 1999. Multiresolution-based image fusion with additive wavelet decomposition. *IEEE Transactions on Geoscience and Remote Sensing*, 37(3):1204–1211. DOI: 10.1109/36.763274. 27

[28] Vivone, G., Restaino, R., Dalla Mura, M., Licciardi, G., and Chanussot, J. 2014. Contrast and error-based fusion schemes for multispectral image pan-sharpening. *IEEE Geoscience and Remote Sensing Letters*, 11(5):930–934. DOI: 10.1109/lgrs.2013.2281996. 27

[29] Otazu, X., González-Audícana, M., Fors, O., and Núñez, J. 2005. Introduction of sensor spectral response into image fusion methods. Application to wavelet-based methods. *IEEE Transactions on Geoscience and Remote Sensing*, 43(10):2376–2385. DOI: 10.1109/tgrs.2005.856106. 27

[30] Ulfarsson, M. O., Palsson, F., Mura, M. D., and Sveinsson, J. R. Sentinel-2 sharpening using a reduced-rank method. *IEEE Transactions on Geoscience and Remote Sensing*, to be published. DOI: 10.1109/tgrs.2019.2906048. 27

CHAPTER 4

Deep Learning-Based Image Fusion Approaches in Remote Sensing

In this chapter, recently developed deep learning approaches for remote sensing image fusion are presented. First, different deep learning networks used are described. In addition, their training procedure is described. Then, the way deep learning models are applied to the fusion problem is discussed in detail. To this end, several representative methods are first mentioned.

4.1   TYPICAL DEEP LEARNING MODELS

Deep learning is a subset of machine learning algorithms that use a set of layers of processing elements for nonlinear data processing in a supervised or an unsupervised manner. Front layers of a deep learning model extract low-level features of input image data while deeper layers extract high-level features. Backend layers of a deep learning model perform classification or regression similar to classical pattern recognition. In other words, both feature extraction and pattern recognition are done together in a deep learning model while in conventional machine learning, feature extraction and classification are done separately.

In general, three main deep learning architectures are considered:

- generative,

- discriminative, and

- hybrid.

Pre-trained layers are used by some of the layers of several deep learning solutions. This so-called transfer learning approach reduces the challenge of learning low-level hidden layers. Each layer may be pre-trained (DNNs) and later included in fine-tuning steps in the overall model. Discriminative deep neural networks achieve differential computing capacity by stacking the output of each layer with the original data or using a number of combinations. Such deep learning models consider the outputs as a conditional distribution over all possible label sequences for a given input sequence, which is then optimized based on an objective function. Hybrid deep learning models combine the properties of the generative and discriminative architectures.

Denoising Autoencoder (DA):   Autoencoder is a deep learning model that is able to learn a variety of coded sequences. A simple type of autoencoder is a multi-layer perceptron that comprises an input layer, one or more hidden (representation) layers and an output layer. The key distinction between the autoencoder and the multi-layer perceptron is the number of nodes on the output layer. In the case of an autoencoder, the output layer comprises the same number of nodes as the input layer. Instead of estimating target values as an output vector, the autoencoder estimates its input. If the number of nodes in the hidden layers is smaller than the input/output nodes, the activation of the last hidden layer is called a compressed approximation of the inputs. Conversely, where the hidden layer nodes are greater than the input layer, the autoencoder will theoretically learn the identity feature and lose its effectiveness. A Denoising Autoencoder (DA) architecture [1] is described in this part, which is among the earliest deep learning models used in remote sensing.

Consider a vectored input data represented by $x \in [0, 1]^d$, which is mapped to $y \in [0, 1]^{d'}$ using a deterministic mapping $y = f_\theta(x) = s(\mathbf{W}x + b)$ in which $\theta = \{\mathbf{W}, b\}$ are the mapping parameters with $\mathbf{W}$ denoting a $d \times d'$ weight matrix and b a bias. The elements of y are mapped back to a reconstructed vector represented by $z \in [0, 1]^d$. The mapping function for the reconstruction is expressed by $z = g_{\theta'}(y) = s(\mathbf{W}'y + b')$. Here, each training frame $x^{(i)}$ is mapped to a hidden representation frame $y^{(i)}$ which is then mapped back to a reconstructed frame $z^{(i)}$. The parameters of the network are normally obtained via solving the following optimization problem:

$$\theta, \theta' = \arg\min_{\theta, \theta'} \frac{1}{n} \sum_{i=1}^{n} L\left(x^{(i)}, z^{(i)}\right), \tag{4.1}$$

where $L(\cdot)$ denotes the l_2-norm $\left\|x^{(i)} - z^{(i)}\right\|_2^2$.

The problem with DA is that in some applications during training, the weights may vary considerably which leads to a problem known as gradient vanishing. This problem is related to the structure of the data. One way to avoid the gradient vanishing is via weight regularization by restricting the range of weight updates to a prescribed amount. Equation (4.1) is thus modified as follows:

$$\theta, \theta' = \arg\min_{\theta, \theta'} \frac{1}{n} \sum_{i=1}^{n} L\left(x^{(i)}, z^{(i)}\right) + \frac{\lambda}{2} \left(\|\mathbf{W}\| + \|\mathbf{W}'\|\right), \tag{4.2}$$

where λ indicates a constant as a tuning parameter. It is worth mentioning that in practice it is seen that during training phase, only a portion of the weights of each layer remains active. This is important from the complexity point of view. This issue can be taken into consideration in Eq. (4.2) by using a sparsity constraint as follows:

$$\theta, \theta' = \arg\min_{\theta, \theta'} \frac{1}{n} \sum_{i=1}^{n} L\left(x^{(i)}, z^{(i)}\right) + \frac{\lambda}{2} \left(\|\mathbf{W}\| + \|\mathbf{W}'\|\right)$$
$$+ \beta KL(\rho\|\widehat{\rho}). \tag{4.3}$$

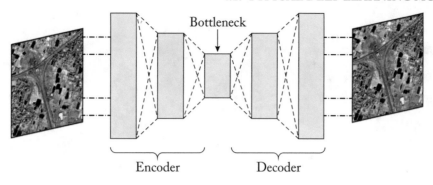

Figure 4.1: Typical DA architecture.

The last term is defined as:

$$KL\left(\rho\|\widehat{\rho}\right) = \rho \log \frac{\rho}{\widehat{\rho}} + (1 - \rho) \log \frac{1 - \rho}{1 - \widehat{\rho}} \tag{4.4}$$

and $\widehat{\rho} = \frac{1}{n} \sum_{i=1}^{n} s(x^{(i)})$ is the average activation of a hidden layer. A typical architecture of DA is shown in Figure 4.1.

Convolutional Neural Network (CNN):   A Convolutional Neural Network (CNN) is the most widely used deep learning model. CNNs are now extensively used in various computer vision applications such as self-driving cars, gesture recognition, and automatic plate recognition. CNN consists of one or more convolution layers followed by one or more fully connected layers as in a generic multi-layer neural network. The CNN architecture is designed to take advantage of the 2D form of the input image (or other 2D input data such as spectrogram). This is accomplished by local connections and fixed weights followed by some sort of pooling layers resulting in translational invariant features. Another advantage of CNNs is that they are faster to train and have less parameters than fully connected models with the same number of hidden units.

A CNN consists of a number of convolutional and optional subsampling layers followed by fully connected layers. A typical architecture of CNN consisting of several convolution and pooling layers is shown in Figure 4.2.

Generative Adversarial Networks (GANs):   The use of Generative Adversarial Networks (GANs) has been steadily growing in DNN [2] and has demonstrated to provide significant performance improvement. Training a GAN architecture involves a "generator" and a "discriminator" networks together, where the first network synthesizes realistic images at input, and the second architecture classifies input data as synthetic or real.

Originally, in the GAN architecture [2], the generator is fed with randomized noise input which leads to various outputs, based on statistical properties. For image enhancement purposes, a specific GAN architecture, known as the conditional GANs (cGANs), is more suitable because

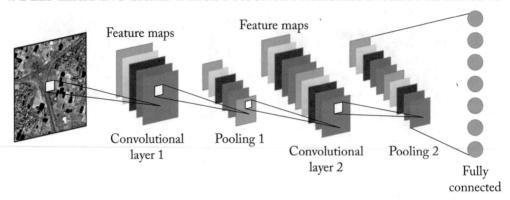

Figure 4.2: Typical CNN architecture.

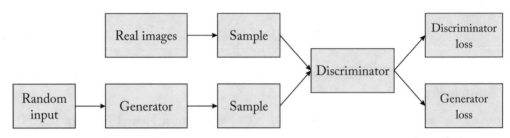

Figure 4.3: Typical GAN architecture.

the input to the generator is the image itself, while it may be somewhat different from the output, such as the edge map [3]. The standard GAN architecture is illustrated in Figure 4.3.

The training of a GAN model involves optimizing a min-max problem in which the goal is to predict weights and biases of both parts of the architecture (generator θ_G and the discriminator θ_D) as follows (in the following notation $\mathbf{I}^R$ is the real image and $\mathbf{I}^{IN}$ is the low-quality input image):

$$\min_{\theta_G} \max_{\theta_D} \mathbb{E}\left[\log D_{\theta_D}\left(\mathbf{I}^R\right)\right] + \mathbb{E}\left[\log\left(1 - D_{\theta_D}\left(G_{\theta_G}\left(\mathbf{I}^{IN}\right)\right)\right)\right] . \qquad (4.5)$$

An important work that shows the abilities of GAN in typical inverse imaging problems is the super-resolution GAN (SRGAN) architecture [4]. The advantages of GANs are summarized below.

- **GANs involve unsupervised learning**: Obtaining labeled data is a time-consuming process. GANs do not require labels for training. They utilize internal representations of data, that is they are trained with unlabeled data.

- **GANs are data generators**: A key feature of GANs is that they are capable of generating data very similar to real data. Generated data by GANs such as text, images, and videos are hard to distinguish from real data.

- **GANs can learn distributions of data**: Since GANs can learn internal representations of data, they can extract meaningful features.

- **GANs trained discriminator can serve as a classifier**: The second module for the learning process is discriminator where at the end of the training stage can be considered to be a classifier.

Convolutional AutoEncoder (CAE):   The convolution operator allows filtering input data in order to extract and represent some part of its content. Autoencoders in their conventional form do not take into account the fact that a signal can be seen as a sum of other signals. On the other hand, Convolutional AutoEncoders (CAEs) use the convolution operator to address this aspect. They learn to encode the input as a set of simpler signals and then attempt to reconstruct the input from these simpler signals. CAE addresses the filtering operation task in a different way. This model learns the optimal filters by minimizing the reconstruction error. These filters (kernels) are utilizable in any other computer vision task.

CAE is widely used for unsupervised learning of convolutional operations. When these filters are learned by training, they can be used to provide representative features for any input. They can also be used to have a compact representation of the input for classification or regression.

CAE is a type of CNN with the main difference that CAE is trained end-to-end to learn the filters and the representative features are combined to achieve the classification. However, CNN is trained only to learn the filters for extracting features that enable reconstructing the input.

CAE scales well to realistic-sized high-dimensional images because the number of parameters required to produce a feature map is always the same, no matter the size of the input. Thus, CAE is a *general purpose feature extractor* different from Autoencoders or DAs that do not take into consideration the 2D image structure. In practice, in Autoencoders or DAs, the image must get unrolled into a single vector and the network must be built while satisfying the number of inputs constraint. In other words, Autoencoders introduce redundancy in the parameters, *forcing each feature to be global*, while CAE does not.

Encoding: To follow the structural characteristics of the input image data and to extract high level features across S dimensions $\widetilde{\mathbf{M}} = \{M_1, M_2, \ldots, M_S\}$, n convolution kernels $F^1 = \{F_1^{(1)}, F_2^{(1)}, \ldots, F_n^{(1)}\}$ are considered to create n intermediate features which can be expressed by the following equation:

$$T_m = s\left(\widetilde{\mathbf{M}} \circ F_m^{(1)} + b_m^{(1)}\right), \quad m = 1, 2, \ldots, n, \tag{4.6}$$

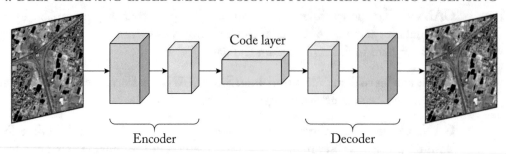

Figure 4.4: Typical CAE architecture.

where s denotes the sigmoid activation function. The intermediate feature maps correspond to the compressed version of the input data. Several convolution layers are normally connected to each other to form a deep model.

Decoding: This part involves reconstructing the input image from its n intermediate feature maps. The reconstructed input image $\widehat{\mathbf{M}}$ is obtained via the convolution of the intermediate feature maps $T = \{T_m\}_{m=1}^{n}$ with the convolutional filters $F^2 = \{F_1^{(2)}, F_2^{(2)}, \ldots, F_n^{(2)}\}$ as follows:

$$\widehat{\mathbf{M}} = s\left(T \circ F_m^{(2)} + b_m^{(2)}\right). \tag{4.7}$$

With the output having the same dimension as the input, any loss function $\mathcal{L}(\cdot)$ such as MSE below can be used for updating the network weights

$$\mathcal{L}\left(\widetilde{\mathbf{M}}, \widehat{\mathbf{M}}\right) = \frac{1}{2}\left\|\widetilde{\mathbf{M}} - \widehat{\mathbf{M}}\right\|_2^2. \tag{4.8}$$

A typical architecture of CAE is shown in Figure 4.4.

4.2   SINGLE-OBJECTIVE LOSS FUNCTION

In this section, a recently developed method that has received much attention is covered.

Denoising Autoencoder modeling of CS-based framework: This method [5] begins by considering the HRPAN image $\mathbf{P}$ and LRMS image $\mathbf{M}$. First, the LRMS image $\widetilde{\mathbf{M}}$ is upsampled to the size of $\mathbf{P}$, generating the co-registered LRMS image $\widetilde{\mathbf{M}}$. Then, $\mathbf{P}$ and each band of $\widetilde{\mathbf{M}}$ is normalized to the range $[0,1]$. Furthermore, the LRPAN image $\mathbf{P}_L$ is computed through a linear summation of LRMS image $\widetilde{\mathbf{M}}$. The HR patches $\{x_{\mathbf{p}}^i\}_{i=1}^{N}$ and the corresponding LR patches $\{y_{\mathbf{p}}^i\}_{i=1}^{N}$ are extracted from $\mathbf{P}$ and $\mathbf{P}_L$, respectively, creating the training set $\{x_{\mathbf{p}}^i, y_{\mathbf{p}}^i\}_{i=1}^{N}$, where N is the number of training image patches. Next, the DNN is trained using the Modified Sparse Denoising Autoencoder (MSDA) and Stacked Modified Sparse Denoising Autoencoder (S-MSDA) via the training set $\{x_{\mathbf{p}}^i, y_{\mathbf{p}}^i\}_{i=1}^{N}$. For pansharpening purposes, this method considers an

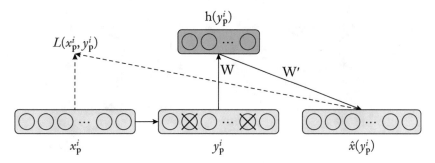

Figure 4.5: Flowchart of Denoising Autoencoder modeling of CS-based framework.

MSDA algorithm for pre-training each layer of the DNN model, which models the highly nonlinear relationship between HR image patches $\{x_{\mathbf{p}}^i\}_{i=1}^N$ (as target data) and the corresponding LR image patches $\{y_{\mathbf{p}}^i\}_{i=1}^N$ (as degraded data); see Figure 4.5.

The pre-training of the MSDA involves computing the value of the vector Θ by optimizing the error function as per the objective loss. More specifically, given the image patch pairs $\{x_{\mathbf{p}}^i, y_{\mathbf{p}}^i\}_{i=1}^N$ of training samples, the parameter Θ is trained based on the steps specified in Figure 4.4.

From a mathematical perspective, let $\{x_{\mathbf{p}}^i\}_{i=1}^N$ be the target data and $\{y_{\mathbf{p}}^i\}_{i=1}^N$ be the degraded input. The forwarding layers of MSDA, containing the encoder and the decoder, can then be represented as follows:

$$h\left(y_{\mathbf{p}}^i\right) = s(W y_{\mathbf{p}}^i + b) \tag{4.9}$$

$$\widehat{x}\left(y_{\mathbf{p}}^i\right) = s(W' h\left(y_{\mathbf{p}}^i\right) + b'). \tag{4.10}$$

Next, the parameter $\Theta = \{W, W', b, b\}$ is optimized by minimizing the following objective function:

$$L_1\left(\{x_{\mathbf{p}}^i, y_{\mathbf{p}}^i\}_{i=1}^N, \Theta\right) = \frac{1}{N}\sum_{i=1}^N L\left(x_{\mathbf{p}}^i, \hat{x}\left(y_{\mathbf{p}}^i\right)\right) + \frac{\lambda}{2}\left(\|W\| + \|W'\|\right)$$
$$+ \beta KL(\rho\|\widehat{\rho}). \tag{4.11}$$

To construct a deep model, a series of MSDAs are concatenated to obtain S-MSDAs. The S-MSDA architecture is illustrated in Figure 4.6. To produce enhanced pansharpening outcomes, after the pre-training step, the trained DNN is fine-tuned by minimizing the following loss function:

$$L_2\left(\{x_{\mathbf{p}}^i, y_{\mathbf{p}}^i\}_{i=1}^N, \Theta\right) = \frac{1}{N}\sum_{i=1}^N L\left(x_{\mathbf{p}}^i, \hat{x}(y_{\mathbf{p}}^i)\right) + \frac{\lambda}{2}\sum_{l=1}^L \|W_l\|_F. \tag{4.12}$$

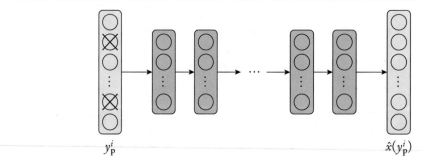

y_p^i $\hat{x}(y_p^i)$

Figure 4.6: Architecture of S-MSDA for network fine-tuning.

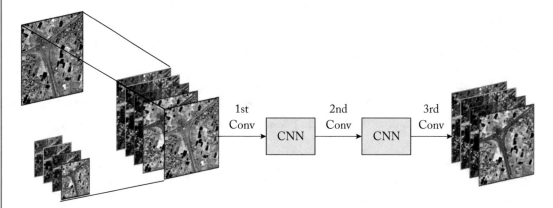

Figure 4.7: CNN architecture proposed for pansharpening using [7].

In which L denotes the total number of MSDAs, and W_l indicates the parameter of the l-th layer in the DNN. Since the sparsity has already been utilized in the pre-training stage, it is removed in this step.

It is anticipated that the relationship between HR/LR PAN image patches is the same as that between HR/LR MS image patches. This is a reasonable assumption based on the spectral signature of the PAN and MS images. The HR MS patches $\hat{x}_k^j$ are obtained based on the feedforward functions in Eqs. (4.9) and (4.10). Therefore, the fused product image $\widehat{\mathbf{M}}$ can be reconstructed by averaging the overlapping image patches x_k^j in all the bands.

Finally, a residual compensation method [6] is used to improve the fusion performance both objectively and visually.

CNN Model for Pansharpening: Figure 4.7 shows a schematic representation of the architecture in [7]. First, the four LRMS bands are up-scaled by factor 4. This way, the network gets trained at the desired resolution without any requirement to down-up-sample steps. The

Table 4.1: Parameters for the CNN model

Layer No.	Type	Description
1st layer	9×9	Kernel size
	ReLU	Activation function
2nd layer	5×5	Kernel size
	ReLU	Activation function
3rd layer	5×5	Kernel size
	Linear Unit	Activation function

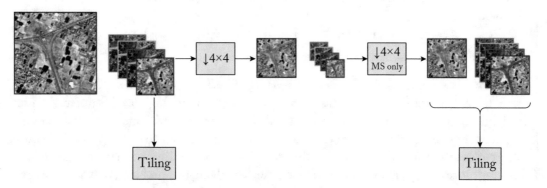

Figure 4.8: Training procedure based on Wald protocol.

outcome consists of only four bands with respect to the PAN resolution version for input MS channels. In order to improve the performance, the 1×1 kernels of the central layer is used. Table 4.1 lists the parameters of the entire model. It is valuable to say that this is quite a simple CNN model, and thus relatively easy to train.

After training and testing the CNN model, the output of the model is a 4-band MS image with the PAN resolution. As shown in Chapter 2, for all practical purposes, the HRMS image does not exist. This creates a variety of issues not only in terms of performance evaluation, but also in terms of the DNN learning process. In this approach, these two issues are discussed using the Wald's protocol, which address operating on a down-sampled MS image for which the intact MS component is used for objective evaluation.

The training procedure based on the Wald's protocol is illustrated in Figure 4.8. During CNN training, the model is trained to generate output tiles that match as closely as possible to the reference tiles. The learned parameters are finally used for the pansharpening of the real MS images at their original resolution. Obviously, the entire procedure relies on the assumption that performance does not depend critically on the resolution scale. Hence, to reduce possible mismatches, the data are smoothed before downsampling using a filter that matches the MTF of the sensor. Also, the MS component is upsmapled using the interpolation filter. In both cases, the results of the fusion have to be assessed not only on the low-resolution images but also on the full-scale images.

The training procedure is done on the basis of backpropagation algorithm and stochastic gradient descent. Updating iteration corresponds to a batch size of 128 input tiles extracted randomly from the training data. The batch MSE between the fused product $\widehat{\mathbf{M}}$ and its reference $\mathbf{M}$ is obtained as follows:

$$L = \frac{1}{N} \sum_{n=1}^{N} \left\| \widehat{\mathbf{M}}_n - \mathbf{M}_n \right\|. \tag{4.13}$$

SGD uses a momentum controlling parameter, hence, at iteration $i + 1$, the update is done as follows:

$$W_{i+1} = W_i + \Delta W_i = W_i + \mu \cdot \Delta W_{i-1} - \alpha \cdot \nabla L_i, \tag{4.14}$$

where μ denotes the momentum, and α a learning rate.

Deep Residual Learning for Pansharpening (DRPNN):   It is seen that a CNN network with more filters and hidden layers results in extracting better useful high level features. As a result, a higher estimation accuracy can be obtained. However, the gradients of the estimation loss to parameters in the shallow hidden layers cannot be passed via back propagation because of the gradient vanishing issue, thus averting the DNN from getting trained well. Deep residual learning procedure [8] is an innovative technique for solving this problem, in which the conventional form of mapping $\widehat{\mathbf{M}} \approx CNN(\widetilde{\mathbf{M}})$ is substituted with $\widehat{\mathbf{M}} - \widetilde{\mathbf{M}} \approx RES(\widetilde{\mathbf{M}})$. Presumably, in the residual image, $\widehat{\mathbf{M}} - \widetilde{\mathbf{M}}$, most pixels are very close to zero, and the spatial distribution of the residual features will exhibit very sparse. Hence, looking for a distribution that is very close to the optimal for $\{W, b\}$ becomes easier, which allows one to add more hidden layers to the network and improve its performance. However, in the pan-sharpening task, the size of the final output $\widehat{\mathbf{M}}(H \times W \times N)$ is not the same as the size of the input $\widetilde{\mathbf{M}}(H \times W \times (N + 1))$. Consequently, rather than predicting the residual features directly, the process through the DRPNN with L layers is divided into the following two stages.

Stage 1: The first to the $(L - 1)$th layers are concatenated under a skip connection to predict the residual between $F^{Stage\ 1}$ and $\widetilde{\mathbf{M}}$. The convolution operations in each layer are obtained as follows:

$$F_0 = \widetilde{\mathbf{M}}, \ F_l = \max\left(0, W_l \circ F_{l-1} + b_l\right), l = 1, \ldots, L - 1. \tag{4.15}$$

The residual output from the $(L - 1)$-th layer is then added to $\widetilde{\mathbf{M}}$ to obtain $F^{Stage\ 1}$,

$$F^{Stage\ 1} = \widetilde{\mathbf{M}} + F_{L-1}. \tag{4.16}$$

Stage 2: The L-th layer of the DRPNN is established to reduce the dimension of the spectral space from $(N + 1)$ bands to N bands via the last 3D convolution operation in the

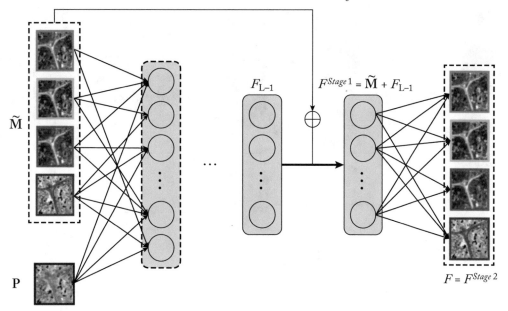

Figure 4.9: Flowchart of deep model training via DRPNN.

network, resulting in the final prediction $F^{Stage\ 2}$ $(H \times W \times N)$:

$$\widehat{\mathbf{M}} = F^{Stage\ 2} = W_L \circ F^{Stage\ 1} + b_L. \tag{4.17}$$

The framework of the DRPNN is depicted in Figure 4.9.

PanNet: The architecture for PanNet is shown in Figure 4.10. PanNet was motivated by considering the spectral preservation and spatial injection in its architecture [9]. A straightforward way of using a DNN model for image fusion in remote sensing can be a network that learns the nonlinear mapping relationship between the inputs $(\mathbf{P}, \mathbf{M})$ and the outputs $\mathbf{X}$ by minimizing the following loss function:

$$\mathcal{L} = \| f_W\ (\mathbf{P},\ \mathbf{M}) - \mathbf{X}\|^2_F\ , \tag{4.18}$$

where f_W represents a neural network and W its parameters. This idea is utilized by the PNN method (described above) [9], which directly inputs $(\mathbf{P}, \mathbf{M})$ into a deep CNN to estimate $\mathbf{X}$. Although this method gives satisfactory performance, it does not address image characteristics.

   Motivated by the success of PNN, the PanNet architecture is built to perform the pansharpening task. As with PNN, PanNet also uses a CNN but its structure differs from PNN [7]. As reported in [7], convolutional filters are useful since they can extract the high correlation across different bands of the multispectral images. PanNet attempts to preserve both spectral and spatial content. Figure 4.11 illustrates a number of PanNet network structures. The first

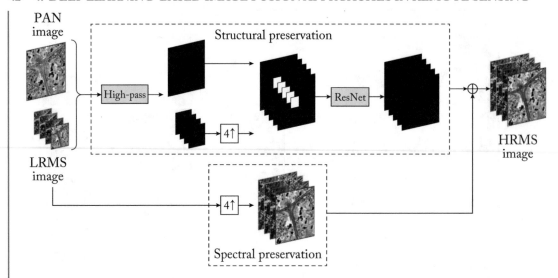

Figure 4.10: Architecture of PanNet.

structure is vanilla ResNet which focuses on spatial information, while the second structure only focuses on spectral information. The third structure takes into consideration both spatial and spectral aspects.

Spectral preservation term: To fuse spectral information, $\mathbf{M}$ is upsampled and a skip connection is added to the network as follows:

$$\widehat{\mathcal{L}} = \| f_W (\mathbf{P}, \mathbf{M}) + \uparrow \mathbf{M} - \mathbf{X} \|_F^2. \tag{4.19}$$

$\uparrow \mathbf{M}$ indicates the upsampled MS image. This term ensures that $\mathbf{X}$ shares the spectral content of $\mathbf{M}$. Unlike variational methods, $\mathbf{X}$ is not convolved with a smoothing kernel, thus allowing the network to correct for high-resolution differences.

Structural preservation term: Several variational methods utilize the high-pass information contained in the PAN image to allow structural consistency in the fused outcome. As a result, high-pass filtering is applied to the PAN and up-sampled MS images and the results are inputted to the network. The following term reflects this modification:

$$\widehat{\mathcal{L}} = \| f_W (G(\mathbf{P}), \uparrow G(\mathbf{M})) + \uparrow \mathbf{M} - \mathbf{X} \|_F^2. \tag{4.20}$$

The high-pass operation is obtained through the G operator. Since $\uparrow \mathbf{M}$ is low resolution, it can be regarded as containing the low-pass spectral content of $\mathbf{X}$, representing the term $\uparrow \mathbf{M} - \mathbf{X}$. This allows the network weights f_W to learn a mapping that fuses the high-pass spatial information contained in the PAN image into $\mathbf{X}$. The term $\uparrow G(\mathbf{M})$ is inputted to the network

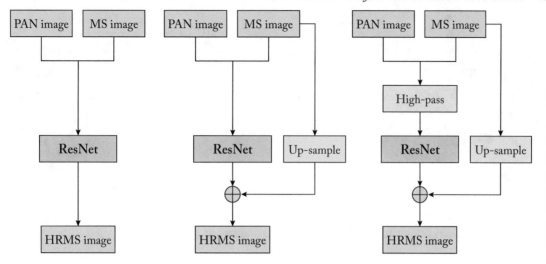

Figure 4.11: Architectures used to test the PanNet model.

in order to learn how the spatial information in the PAN image maps to the spectral bands in **X**.

CAE Training for Pansharpening: A CAE network is used to improve the spatial information of the LRMS bands by learning the nonlinear relationship between a PAN image and its spatially degraded version [10], as illustrated in Figure 4.12. As indicated earlier, a PAN image is considered to be the reference spatial image. A spatially degraded version of the PAN image is first generated using an interpolation filter. The original and degraded PAN images are partitioned into $p \times p$ patches with r overlapping pixels. A CAE network is then used to learn the nonlinear relationship between the original PAN patches and corresponding degraded PAN patches as its target and input, respectively. After training, the CAE network generates approximated high resolution LRMS patches as its output in response to LRMS patches as its input.

The original PAN patches can be stated as $\left\{ \mathbf{P}_i^H \right\}_{i=1}^{C}$, and spatially degraded PAN patches as $\left\{ \widetilde{\mathbf{P}}_i^H \right\}_{i=1}^{C}$, to form the target and the input of the CAE network, respectively, where C indicates the number of patches. The network is effectively designed to learn how to inject the spatial information into the degraded image. At each iteration, the output patches of the CAE network are computed as follows:

$$\left\{ \widehat{\mathbf{P}}_i^H \right\}_{i=1}^{C} = F_{Decode} \left(F_{Encode} \left(\left\{ \mathbf{P}_i^H \right\}_{i=1}^{C} \right) \right), \tag{4.21}$$

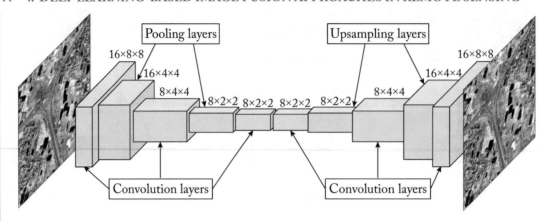

Figure 4.12: CAE architecture used for pansharpening in [10].

where $\left\{\widehat{\mathbf{P}}_i^H\right\}_{i=1}^C$ denotes the output and F_{Encode} and F_{Decode} correspond to the encoding and decoding operations. The weights at each iteration are updated based on the MSE between the original PAN patches and their reconstructed versions:

$$\mathcal{L}\left(\left\{\widehat{\mathbf{P}}_i^H\right\}_{i=1}^C, \{\mathbf{P}_i^H\}_{i=1}^C\right) = \frac{1}{2}\sum_{i=1}^C \left\|\widehat{\mathbf{P}}_i^H - \mathbf{P}_i^H\right\|_2^2. \tag{4.22}$$

The backpropagation algorithm is then employed for training the network. To test the effectiveness of the trained network, the LRMS image is partitioned into patches across N bands, that is $\left\{\widetilde{\mathbf{M}}_{i,j}^L\right\}_{i=1}^C$ for $j = 1, \ldots, N$.

Next, the patch-wised LRMS bands are fed into the CAE trained network. Due to the similarity in the spectral characteristics of PAN and MS images, the trained network is expected to improve the spatial information of the LRMS bands. The input and output of the CAE network are illustrated in Figure 4.13. From this figure, it can be seen that an estimated high-resolution LRMS is obtained at the output of the network for each band. In fact, not only the reconstructed version preserves the spectral information of LRMS bands but also it carries more spatial information in comparison with the input patches. What makes this approach different than the previous approaches is that the estimated high resolution LRMS is used instead of the original LRMS. After tiling the estimated high-resolution LRMS bands, the fusion process is carried out via the following equation:

$$\mathbf{M}_i = \widehat{\mathbf{M}}_i + g_i(\mathbf{P} - \mathbf{I}), \tag{4.23}$$

where $\widehat{\mathbf{M}}_i$ denotes the i-th estimated high-resolution LRMS band obtained from the trained CAE network. Figure 4.14 provides an illustration of the steps involved in this method. As

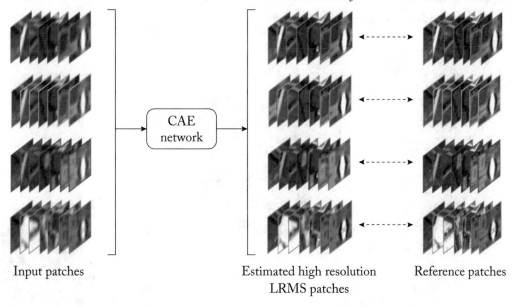

Input patches   Estimated high resolution   Reference patches
                                       LRMS patches

Figure 4.13: Testing phase of the CAE architecture used in [10].

illustrated in this figure, the optimal weights are applied to the LRMS bands in order to obtain an estimation of the low-resolution PAN image. Then, the primitive detail map is obtained via Eq. (3.2). Next, the injection gains noted in Eq. (3.6) are used to acquire the refined detail map. After obtaining and tiling the estimated high resolution LRMS patches, the first term of Eq. (4.22) is computed. The fusion is achieved by combining its two components, as illustrated in Figure 4.14.

Detail-Preserve Cross-Scale Learning for CNN-Based Pansharpening: A very important aspect of deep learning solutions for pansharpening is to what extent the capability exists for generalization from the resolution of training to the resolution of testing. The training set is obtained by a resolution shift (Wald's Protocol [11]). In remote sensing, the problem of shifting from the training resolution to the testing resolution is a crucial issue, because all the sample images are captured at a specified distance from the earth's surface, thereby at a relatively constant distance from the ground sample. Consequently, no representative variations of these elements would be identified in the training dataset, resulting in a misalignment between the training and the test datasets. That is why the large gain of CNN-based methods in comparison with conventional methods in the reduced resolution evaluation structure occurs with a less distinguishable performance in the full-resolution. A number of attempts to address this issue are made in pansharpening [12–14]. The underlying idea in these attempts is to incorporate a complementary loss term that deals with the target-resolution behavior of the model. Based on

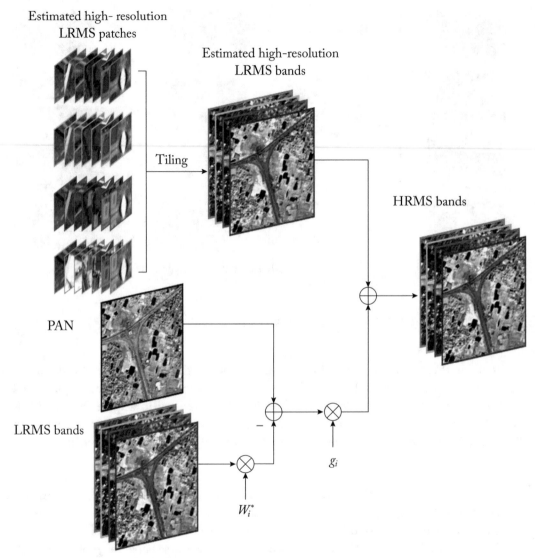

Figure 4.14: Steps involved in the presented pansharpening method in [10].

this idea, a new pansharpening method is developed which is illustrated in Figure 4.15. In particular, this training scheme is applied directly in the fine-tuning stage of the pre-trained model with the parameters Φ_0. This approach takes into consideration the performance loss between the training and test sets [15].

Basically, in the fine-tuning stage, along with the term $\mathcal{L}_{LR}$ defined in the reduced-resolution, a full-resolution term $\mathcal{L}_{HR}$ is considered by appropriately processing the target image

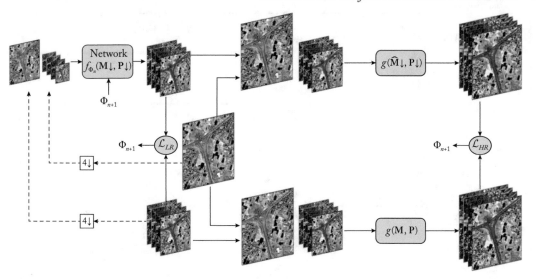

Figure 4.15: Steps involved in detail-preserve cross-scale learning [15].

as indicated in Figure 4.15:

$$\mathcal{L} = \alpha \mathcal{L}_{LR} + \beta \mathcal{L}_{HR}. \tag{4.24}$$

The terms $\mathcal{L}_{LR}$ and $\mathcal{L}_{HR}$ are defined as follows:

$$\mathcal{L}_{LR} = \left\| \widehat{\mathbf{M}} \downarrow - \mathbf{M} \downarrow \right\|_1 = \| f_{\Phi_n} (\mathbf{M} \downarrow, \mathbf{P} \downarrow) - \mathbf{M} \| \tag{4.25}$$

$$\mathcal{L}_{HR} = \left\| g \left(\widehat{\mathbf{M}} \downarrow, \mathbf{P} \right) - g \left(\mathbf{M}, \mathbf{P} \right) \right\|_1$$
$$= \| g \left(f_{\Phi_n} (\mathbf{M} \downarrow, \mathbf{P} \downarrow), \mathbf{P} \right) - g \left(\mathbf{M}, \mathbf{P} \right) \|_1, \tag{4.26}$$

in which the multispectral image fusion at the final resolution is done using a mapping function $g(\cdot, \cdot)$. Specifically, one of the traditional methods that denotes reasonable performance on preserving the spatial consistency is utilized (MTF-GLP-HPM). In fact, this conventional algorithm highly depends on the extracted spatial information from the PAN image while restricting the spectral distortion effect. Therefore, the cross-scale consistency is enforced due to the training on low- and high-resolution data jointly.

Unsupervised Pansharpening using Generative Adversarial Networks: The final target of the pansharpening is to preserve the spatial and spectral information in the final product. However, current CNN-based approaches typically regard pansharpening as a deep learning module. Although PanNet [12] relies on preservation spectral and spatial content, it attains a fused product by combining the up-scaled MS data with the high-frequency information given by CNN that leads to blurred outcomes. In comparison, the above methods depend on the reference image, i.e., the Wald's protocol [11], where all the reference images are degraded by the

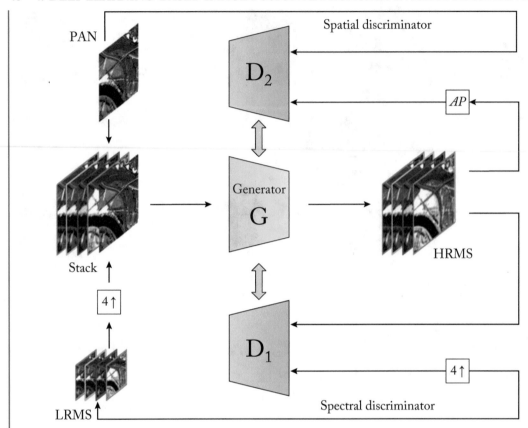

Figure 4.16: Unsupervised deep learning model for pansharpening.

Gaussian kernel and then downscaled by a factor of 4. The rationale behind this work [16] is that the ill-posed pansharpening problem cannot be solved without blurriness by reversing the down and Gaussian kernels when dealing with Wald's protocol. Therefore, an unsupervised pansharpening model is proposed (called Pan-GAN here) in order to utilize the intact source images as the training data to get the HRMS fused image without supervision by the ground-truth data.

First, the original MS data is interpolated to the resolution of the PAN image and then stacked to from the five-channel data to be fed into the generator G to obtain the HRMS image at the output. It is worth mentioning that the first channel is considered to be the PAN image. It is found that without considering two discriminators for the GAN, the final HRMS suffers from either severe spectral distortion or spatial inconsistencies.

Two discriminators are used in this model for preserving both spectral and spatial information. The flowchart of the developed unsupervised method is shown in Figure 4.16. In order to deal with the preservation of spectral and spatial information, two tasks using two dis-

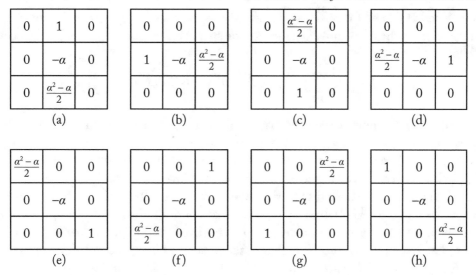

Figure 4.17: Eight directional masks for extraction of the edge map: (a) $\mathbf{x}^+$, (b) $\mathbf{y}^+$, (c) $\mathbf{x}^-$, (d) $\mathbf{y}^-$, (e) LUD, (f) LLD, (g) RUD, and (h) RLD.

criminators are considered simultaneously. The first discriminator labeled as D_1 in Figure 4.16 corresponds to the spectral discriminator, and its aim is to check the spectral consistency of the generated image with the LRMS image data. Another discriminator is considered to preserve the spatial content in the HRMS. In fact, the aim of the second discriminator D_2 is to check the spatial consistency of the generated image with the PAN image. Since the PAN image is a single-band data, the average pooling operation is applied to the generated image to obtain a single-band data.

4.3   MULTI-OBJECTIVE LOSS FUNCTION

This section includes the latest work on the formulation of the fusion problem as a multi-objective loss function represented by a DNN, in which both the spectral and spatial distortions are minimized. Two terms are added to the conventional loss function (MSE). One term reflects the spatial information and it consists of two parts: (1) the edge information of spectral bands using the fractional-order differentiation masks shown in Figure 4.17 and (2) Q-index [17]. To compute Q-index, first Universal Image Quality Index (UIQI) is obtained as follows (defined earlier in Chapter 2):

$$UIQI\left(\mathbf{A},\mathbf{B}\right)=\frac{\sigma_{AB}}{\sigma_A\sigma_B}\frac{2\mu_A\mu_B}{\left(\mu_A^2+\mu_B^2\right)}\frac{2\sigma_A\sigma_B}{\left(\sigma_A^2+\sigma_B^2\right)}, \tag{4.27}$$

where $\mathbf{A}$ and $\mathbf{B}$ denote the reference and fused images, respectively, $\sigma_{\mathbf{AB}}$ is the covariance of $\mathbf{A}$ and $\mathbf{B}$, and $\mu_{\mathbf{A}}$ is the mean of $\mathbf{A}$. Q-index for the spatial information (Q_s) [17] is then computed as follows:

$$Q_s = \frac{1}{N} \sum_{k=1}^{N} \left| UIQI\left(\mathbf{z}^{(k)}, \mathbf{D}_k^{\mathbf{P}}\right) - UIQI\left(\widetilde{\mathbf{M}}_k, \mathbf{D}_k^{\mathbf{P}^{LP}}\right) \right|^q, \qquad (4.28)$$

where $\mathbf{D}_k^{\mathbf{P}}$ and $\mathbf{D}_k^{\mathbf{P}^{LP}}$ indicate the edge maps of the k-th spectral band extracted using the superimposed mask from the original and low-resolution PAN images, respectively, and $\widetilde{\mathbf{M}}_k$ represents the k-th LRMS band in a frame-based manner. This objective function first computes the UIQI index between the extracted edge maps and the low-resolution PAN image frame. Since the PAN frames are considered to be the target frames, the UIQI index between the output of the network and the PAN frames are computed and optimized with respect to the UIQI index of the input frame. This process allows a more effective injection of the spatial details and thus leads to less spatial distortion. Note that different spectral bands do not share the same detail map. A depiction of the spatial term in the loss function is shown in Figure 4.18.

Next, a second spectral term is added to minimize the spectral distortion in the fused image. The UIQI index of the input frames are first computed in band-wise manner. Then, the band-wise UIQI index of the output frames is similarly computed and compared to the corresponding UIQI index of the input frames. Mathematically, this process can be stated as follows [18]:

$$Q_\lambda = \frac{1}{(N)(N-1)} \sum_{k=1}^{N} \sum_{j=1, k \neq j}^{N} \left| UIQI\left(\mathbf{z}^{(k)}, \mathbf{z}^{(j)}\right) - UIQI\left(\widetilde{\mathbf{M}}_k, \widetilde{\mathbf{M}}_j\right) \right|^P. \qquad (4.29)$$

In Eq. (4.29), the difference between the UIQI terms allows the preservation of the color information in the fused image. A flowchart of the computation of the spectral term is depicted in Figure 4.19.

As described above, the multi-objective method adds two more terms to the conventional or commonly used MSE loss function to more effectively preserve the spectral information and also to more effectively inject the spatial detail. The added terms lead to a more effective optimization of the spectral and spatial distortions. The following fusion framework can be used to obtain the HRMS image:

$$\widehat{\mathbf{M}}_k = \widetilde{\mathbf{M}}_k + f_k(x, \theta), \qquad (4.30)$$

where the term $f_k(x, \theta)$ is the output of the DA network corresponding to the missing detail of the k-th spectral band, named residual in the literature. The new terms are incorporated here into $f_k(x, \theta)$ to form this multi-objective loss function:

$$L_{tot} = \frac{1}{N} \sum_{k=1}^{N} \left\| \mathbf{z}^{(k)} - \widetilde{\mathbf{M}}_k \right\| + \alpha Q_s + \beta Q_\lambda. \qquad (4.31)$$

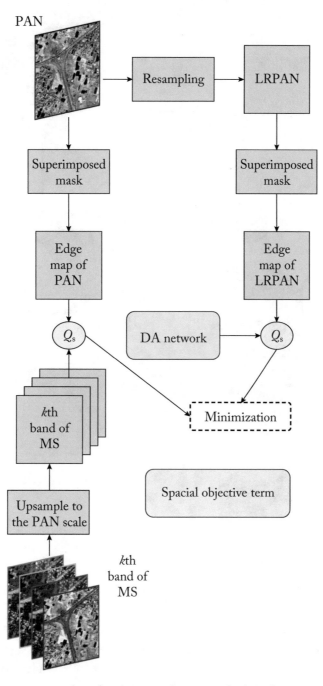

Figure 4.18: Computation pipeline for the spatial term in the loss function.

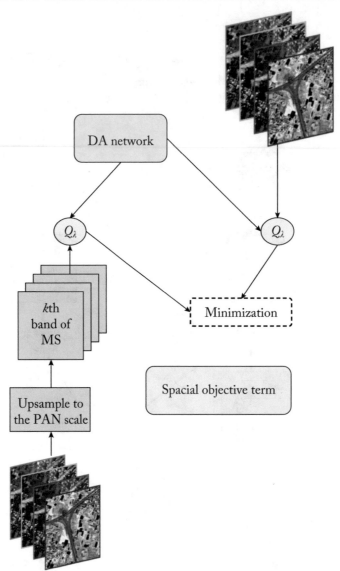

Figure 4.19: Computation pipeline for the spectral term in the loss function.

The weighting factors α and β in Eq. (4.31) can be set to 1 for simplicity. It is worth noting that for optimizing the MSE term in Eq. (4.31), the input is normally selected to be the intensity components of the spectral bands and the output to be the upsampled MS image. In fact, this minimization is similar to the one reported in [16].

4.4 REFERENCES

[1] Vincent, P., Larochelle, H., Bengio, Y., and Manzagol, P. A. 2008. Extracting and composing robust features with denoising autoencoders. *Proc. of the 25th International Conference on Machine Learning*, pages 1096–1103. DOI: 10.1145/1390156.1390294. 32

[2] Goodfellow, I., Pouget-Abadie, J., Mirza, M., Xu, B., Warde-Farley, D., Ozair, S., Courville, A., and Bengio, Y. 2014. Generative adversarial nets. *Advances in Neural Information Processing Systems*, pages 2672–2680. 33

[3] Mirza, M. and Osindero, S. 2014. Conditional generative adversarial nets. *ArXiv Preprint ArXiv:1411.1784*. 34

[4] Ledig, C., Theis, L., Huszár, F., Caballero, J., Cunningham, A., Acosta, A., Aitken, A., Tejani, A., Totz, J., Wang, Z., and Shi, W. 2017. Photo-realistic single image super-resolution using a generative adversarial network. *Proc. of the IEEE Conference on Computer Vision and Pattern Recognition*, pages 4681–4690. DOI: 10.1109/cvpr.2017.19. 34

[5] Huang, W., Xiao, L., Wei, Z., Liu, H., and Tang, S. 2015. A new pan-sharpening method with deep neural networks. *IEEE Geoscience and Remote Sensing Letters*, 12(5):1037–1041. DOI: 10.1109/lgrs.2014.2376034. 36

[6] Yang, S., Wang, M., Chen, Y., and Sun, Y. 2012. Single-image super-resolution reconstruction via learned geometric dictionaries and clustered sparse coding. *IEEE Transactions on Image Process.*, 21(9):4016–4028. DOI: 10.1109/tip.2012.2201491. 38

[7] Masi, G., Cozzolino, D., Verdoliva, L., and Scarpa, G. 2016. Pansharpening by convolutional neural networks. *Remote Sensing*, 8(7):594. DOI: 10.3390/rs8070594. 38, 41

[8] Wei, Y., Yuan, Q., Shen, H., and Zhang, L. 2017. Boosting the accuracy of multispectral image pan-sharpening by learning a deep residual network. *IEEE Geoscience and Remote Sensing Letters*, 14(10):1795–1799. DOI: 10.1109/lgrs.2017.2736020. 40

[9] Yang, J., Fu, X., Hu, Y., Huang, Y., Ding, X., and Paisley, J. 2017. PanNet: A deep network architecture for pan-sharpening. *Proc. of the IEEE International Conference on Computer Vision*, pages 5449–5457. DOI: 10.1109/iccv.2017.193. 41

[10] Azarang, A., Manoochehri, H. E., and Kehtarnavaz, N. 2019. Convolutional autoencoder-based multispectral image fusion. *IEEE Access*, 7:35673–35683. DOI: 10.1109/access.2019.2905511. 43, 44, 45, 46

[11] Wald, L., et al. 1997. Fusion of satellite images of different spatial resolutions: Assessing the quality of resulting images. *Photogrammetric Engineering and Remote Sensing*, 63(6):691–699. 45, 47

[12] Yang, J., Fu, X., Hu, Y., Huang, Y., Ding, X., Paisley, J. 2007. PanNet: A deep network architecture for pan-sharpening. *Proc. of the IEEE International Conference on Computer Vision (ICCV)*, Venice, Italy, October 22–29. DOI: 10.1109/iccv.2017.193. 45, 47

[13] Wei, Y. and Yuan, Q. 2017. Deep residual learning for remote sensed imagery pan-sharpening. *Proc. of the International Workshop on Remote Sensing with Intelligent Processing (RSIP)*, Shanghai, China, May 18–21, pages 1–4. DOI: 10.1109/rsip.2017.7958794. 45

[14] Azarang, A. and Ghassemian, H. 2017. A new pan-sharpening method using multi resolution analysis framework and deep neural networks. *Proc. of the 3rd International Conference on Pattern Recognition and Image Analysis (IPRIA)*, Shahrekord, Iran, April 19–20, pages 1–6. DOI: 10.1109/pria.2017.7983017. 45

[15] Vitale, S. and Scarpa, G. 2020. A detail-preserving cross-scale learning strategy for CNN-based pan-sharpening. *Remote Sensing*, 12(3):348. DOI: 10.3390/rs12030348. 46, 47

[16] Ma, J., Yu, W., Chen, C., Liang, P., Guo, X., and Jiang, J. 2020. Pan-GAN: An unsupervised learning method for pan-sharpening in remote sensing image fusion using a generative adversarial network. *Information Fusion*. DOI: 10.1016/j.inffus.2020.04.006. 48, 52

[17] Alparone, L., et al. 2008. Multispectral and panchromatic data fusion assessment without reference. *Photogrammetric Engineering and Remote Sensing*, 74(2):193–200. DOI: 10.14358/pers.74.2.193. 49, 50

[18] Azarang, A. and Kehtarnavaz, N. 2020. Image fusion in remote sensing by multi-objective deep learning. *International Journal of Remote Sensing*, 41(24):9507–9524. DOI: 10.1080/01431161.2020.1800126. 50

CHAPTER 5

Unsupervised Generative Model for Pansharpening

So far, different aspects of conventional and deep learning-based solutions for multispectral image fusion have been considered. Recently, several attempts have been made to enhance the fusion products in terms of cross-scale learning process and new loss functions. In this section, these two issues are discussed for multispectral image fusion.

As discussed in detail in Chapter 4, deep learning models have been applied to multispectral image fusion generating better outcomes than conventional methods. An initial attempt was made in [1] by solving the pansharpening problem via a DNN framework where the non-linear relationship between the low-resolution and high-resolution images was formulated as a denoising autoencoder. In [2], a three-layer CNN was designed to turn the MS image fusion problem into a super-resolution problem. The concept of residual learning in MS image fusion was first introduced in [3], where a deep CNN was used. In [4], a deep denoising auto-encoder approach was utilized to model the relationship between a low-pass version of the PAN image and its high-resolution version. In [5], a deep CNN was developed for image super-resolution (known as SRCNN), which showed superior performance compared to several other methods. In [6], a pansharpening method was introduced by using the SRCNN as a pre-processing step. In [7], a network structure (known as PanNet) was developed by incorporating the prior knowledge of pansharpening toward enhancing the generalization capability and performance of the network. A GAN method (known as PSGAN) was discussed in [8] by minimizing the loss between the generative and discriminator parts. One of the advantages of GANs is that it reduces the blurriness on the fused image. Not only it attempts to decrease the L_1 loss associated with each pixel, but it also attempts to minimize the loss across the entire fused image.

As far as the loss function in DNNs is concerned, a new perceptual loss was presented in [9] to better preserve the spectral information in fused images. In [10], a number of objective functions were examined. In the recently developed deep learning-based methods, the focus is placed on the preservation of spatial details. For example, the CNN model in [11] was designed for preserving details via a cross-scale learning process. To address the effect of gradient vanishing, the concept of dense connection to pansharpening was extended in [12]. Most of the recently developed deep learning-based methods simply train and regularize the parameters of a network by minimizing a spectral loss between the network output and a pseudo Ground Truth (GT) image. As described in Chapter 4, the methods mentioned above primarily use a single objective learning to optimize network parameters and generalize its capability. However, other metrics that can represent both modalities (spatial and spectral) have recently gained more attention. For instance, in [13], based on the correlation maps between MS target images and PAN input images, a loss function was designed to minimize the artifacts of fused images. Also, in [14], it was shown that although a linear combination of MS bands could be estimated from the PAN image, a rather large difference in luminance was resulted. Thus, certain objects could not be differentiated properly. To address this issue, a color-aware perceptual (CAP) loss was designed to obtain the features of a pre-trained VGG network that were more sensitive to spatial details and less sensitive to color differences. The aforementioned methods rely on the availability of GT data for regularizing the network parameters. However, in practice, such data are not available [15].

This chapter examines ways to ease the above two limitations of the existing deep learning models for remote sensing image fusion. The first limitation involves dependency of GT data toward training and regularizing a network and the second limitation involves the use of a generic loss function for parameter estimation. The first limitation is eased by an unsupervised learning strategy based on generative adversarial networks. What is meant by unsupervised learning here is that the label/reference target is not available for training the deep model. A key point for unsupervised learning is that while the data passed through the deep model are abundant, the targets and labels are quite sparse or even non-existent. The second limitation is eased by designing a multi-objective loss function to reflect both spatial and spectral attributes at the same time.

5.1   METHODOLOGY

This section provides a description of our developed unsupervised generative model. To set the stage, let us begin with the general framework of CS methods as covered in Chapter 2. The CS framework can be mathematically expressed by the following equation:

$$\widehat{\mathbf{M}}_k = \widetilde{\mathbf{M}}_k + g_k(\mathbf{P} - \mathbf{I}_k), \tag{5.1}$$

where $\widehat{\mathbf{M}}_k$ and $\widetilde{\mathbf{M}}_k$ denote the high-resolution and upsampled low-resolution MS images, respectively, g_k's are injection gains for spectral bands, $\mathbf{P}$ denotes the PAN image, and $\mathbf{I}_k$ is the k-th intensity component defined as:

$$\mathbf{I}_k = F\left(\widetilde{\mathbf{M}}_k\right), \tag{5.2}$$

where $F(\cdot)$ is a linear/nonlinear combination of spectral bands [16].

5.2   LEARNING PROCESS AND LOSS FUNCTIONS

Given a set of unlabeled information at the input, the purpose of a generative model is to estimate the data distribution. This is a challenging task and finding such distribution could be so time-consuming. Recently, the GANs were developed to estimate the underlying distribution of an unlabeled data. The general architecture of the GANs were described in Chapter 4. Here, the focus is specifically placed on its application in multispectral image fusion.

The aim of pansharpening is to obtain a HRMS image $\widehat{\mathbf{M}}$ using two input data LRMS image $\widetilde{\mathbf{M}}$ and HRPAN image $\mathbf{P}$. The optimization algorithm which is a min-max problem is generally formulated as follows:

$$\widehat{\mathbf{M}} = f\left(\widetilde{\mathbf{M}}, \mathbf{P}; \Theta\right), \qquad (5.3)$$

where $f(\cdot)$ is the pansharpening model that takes the $\widetilde{\mathbf{M}}$ and $\mathbf{P}$ as input and the parameters to be optimized are represented as Θ. The learning process includes two major branches, e.g., generation and discrimination. The first part of the branch is dedicated to generating a set of images from a random distribution and the role of the second branch is to decide whether the generated data is real or fake by comparing it with the available LRPAN image $\mathbf{P}$ and LRMS image $\widetilde{\mathbf{M}}$.

The procedure of training scheme for this method consists of two parts: (1) spectral preservation and (2) spatial preservation. In what follows an explanation of the loss function design and the parameters to optimize is provided.

Spectral Preservation Learning Process: For minimizing the spectral distortion in the fused image, a spectral metric is used to deal with spectral consistency. For this purpose, a discriminator for the learning process is considered, named spectral discriminator here. The MS image data at the original resolution are used as the input of this discriminator. Initially, the output of the generator is inputted to the spectral discriminator. The following objective function is then used to minimize the spectral distortion of the fused image:

$$\mathcal{L}_1 = \mathbb{Q}\left(\widehat{\mathbf{M}}_k^E, \mathbf{M}_k\right), \qquad (5.4)$$

where $\mathbb{Q}(\cdot, \cdot)$ is the UIQI as described in [16], and $\widehat{\mathbf{M}}_k^E$ and $\mathbf{M}_k$ are the estimated high-resolution MS image at the output of the generator and the MS input image at the original resolution, respectively.

Spatial Preservation Learning Process: Another discriminator is considered for the minimization of spatial distortion, named spatial discriminator here. To inject spatial details into the fused image and by noting that the PAN image denotes the reference spatial information, the PAN image at the original resolution is used as the input to the discriminator. The following loss function is then used during the training phase of the generative model:

$$\mathcal{L}_2 = \mathbb{Q}\left(\widehat{\mathbf{I}}_k^E, \mathbf{P}_k\right), \qquad (5.5)$$

where $\widehat{\mathbf{I}}_k^E$ is the linear combination of estimated high-resolution MS images at the output of the generator and $\mathbf{P}_k$ is the histogram matched PAN image with respect to the k-th spectral band. The learning process of the developed method is illustrated in Figure 5.1.

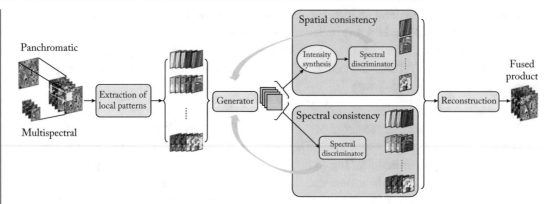

Figure 5.1: Flowchart of the deep generative model for pansharpening.

5.3   REFERENCES

[1] Huang, W., Xiao, L., Wei, Z., Liu, H., and Tang, S. 2015. A new pan-sharpening method with deep neural networks. *IEEE Geoscience and Remote Sensing Letters*, 12(5):1037–1041. DOI: 10.1109/lgrs.2014.2376034. 55

[2] Masi, G., Cozzolino, D., Verdoliva, L., and Scarpa, G. 2016. Pansharpening by convolutional neural networks. *Remote Sensing*, 8(7):594. DOI: 10.3390/rs8070594. 55

[3] Wei, Y., Yuan, Q., Shen, H., and Zhang, L. 2017. Boosting the accuracy of multispectral image pan-sharpening by learning a deep residual network. *IEEE Geoscience and Remote Sensing Letters*, 14(10):1795–1799. DOI: 10.1109/lgrs.2017.2736020. 55

[4] Azarang, A. and Ghassemian, H. 2017. A new pan-sharpening method using multi resolution analysis framework and deep neural networks. *Proc. of the 3rd International Conference on Pattern Recognition and Image Analysis (IPRIA)*, Shahrekord, Iran, April 19–20, pages 1–6. DOI: 10.1109/pria.2017.7983017. 55

[5] Yang, S., Wang, M., Chen, Y., and Sun, Y. 2012. Single-image super-resolution reconstruction via learned geometric dictionaries and clustered sparse coding. *IEEE Transactions on Image Processing*, 21(9):4016–4028. DOI: 10.1109/tip.2012.2201491. 55

[6] Wei, Y. and Yuan, Q. 2017. Deep residual learning for remote sensed imagery pan-sharpening. *Proc. of the International Workshop on Remote Sensing with Intelligent Processing (RSIP)*, Shanghai, China, May 18–21, pages 1–4. DOI: 10.1109/rsip.2017.7958794. 55

[7] Yang, J., Fu, X., Hu, Y., Huang, Y., Ding, X., and Paisley, J. 2017. PanNet: A deep network architecture for pan-sharpening. *Proc. of the IEEE International Conference on Computer Vision*, pages 5449–5457. DOI: 10.1109/iccv.2017.193. 55

[8] Liu, X., Wang, Y., and Liu, Q. 2018. PSGAN: A generative adversarial network for remote sensing image pan-sharpening. *25th IEEE International Conference on Image Processing (ICIP)*, pages 873–877. DOI: 10.1109/icip.2018.8451049. 55

[9] Vitale, S. 2019. A CNN-based pan-sharpening method with perceptual loss. *IGARSS, IEEE International Geoscience and Remote Sensing Symposium*, Yokohama, Japan, pages 3105–3108. DOI: 10.1109/igarss.2019.8900390. 55

[10] Scarpa, G., Vitale, S., and Cozzolino, D. 2018. Target-adaptive CNN-based pan-sharpening. *IEEE Transactions on Geoscience and Remote Sensing*, 56(9):5443–5457. DOI: 10.1109/tgrs.2018.2817393. 55

[11] Vitale, S. and Scarpa, G. 2020. A detail-preserving cross-scale learning strategy for CNN-based pan-sharpening. *Remote Sensing*, 12(3):348. DOI: 10.3390/rs12030348. 55

[12] Wang, D., Li, Y., Ma, L., Bai, Z., and Chan, J. C. W. 2019. Going deeper with densely connected convolutional neural networks for multispectral pan-sharpening. *Remote Sensing*, 11(22):2608. DOI: 10.3390/rs11222608. 55

[13] Bello, J. L. G., Seo, S., and Kim, M. 2020. Pan-sharpening with color-aware perceptual loss and guided re-colorization. *ArXiv Preprint ArXiv:2006.16583*. DOI: 10.1109/icip40778.2020.9190785. 56

[14] Bello, J. L. G., Seo, S., and Kim, M. 2020. Pan-sharpening with color-aware perceptual loss and guided re-colorization. *IEEE International Conference on Image Processing (ICIP)*, pages 908–912. DOI: 10.1109/icip40778.2020.9190785. 56

[15] Ma, J., Yu, W., Chen, C., Liang, P., Guo, X., and Jiang, J. 2020. Pan-GAN: An unsupervised pan-sharpening method for remote sensing image fusion. *Information Fusion*, 62:110–120. DOI: 10.1016/j.inffus.2020.04.006. 56

[16] Azarang, A. and Kehtarnavaz, N. 2020. Multispectral image fusion based on map estimation with improved detail. *Remote Sensing Letters*, 11(8):797–806. DOI: 10.1080/2150704x.2020.1773004. 56, 57

CHAPTER 6

Experimental Studies

As noted in the previous chapters, the image fusion or pansharpening methods can be categorized into two main groups, i.e., conventional methods and deep learning-based methods. It has been shown that deep learning-based methods are more effective than conventional methods. The main reason is the capability of deep learning models to achieve more effective image restoration.

In this chapter, a number of widely used pansharpening methods are selected to conduct a reduced- and full-resolution analysis on different image data. The details on the satellite data used is stated first. To conduct the experiments in a fair manner, the methods whose codes are publicly available are used. To make the comparison more complete, some of the conventional methods described in Chapter 3 are also used in the experiments. After reporting the objective assessments, a discussion of the results is stated from a visual perspective.

6.1   DATASET USED

To conduct both objective and subjective evaluations, three public domain datasets are considered. The sensors related to the datasets are QuickBird, Pleiades-1A, and GeoEye-1. The specifications of these satellites' sensors were described in Chapter 2. The QuickBird data used for the experiments are captured from a jungle area. The geographical area of the data is located at Sundarbans, Bangladesh. A sample MS-PAN pair data for this sensor is shown in Figure 6.1.

For the Pleiades-1A data and in order to cover a wide range of vegetation index, an urban area is selected to conduct the experiments. This data is captured from Paris, France. A sample MS-PAN pair data for this sensor is shown in Figure 6.2. It is worth noting that the original data are very large in terms of pixel dimension. Hence, different regions of the original dataset are cropped to the same size for the experiments.

And finally, for the GeoEye-1 sensor, a coastal region is selected where the data correspond to the Washington area in the U.S. A sample MS-PAN pair of the data is shown in Figure 6.3. As mentioned above, different regions are selected for the experiments reported here. For the QuickBird, Pleiades-1A, and GeoEye-1 sensors, 100, 60, and 50 pairs of MS-PAN images are considered, respectively. The sizes of the PAN and MS images in these datasets are 1024×1024 pixels and 256×256 pixels, respectively.

Figure 6.1: Sample QuickBird imagery data.

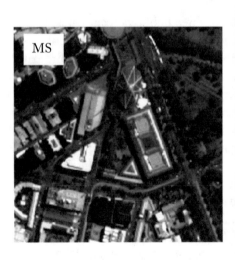

Figure 6.2: Sample Pleiades-1A imagery data.

Figure 6.3: Sample GeoEye-1 imagery data.

6.2   OBJECTIVE ASSESSMENT OF FUSION RESULTS

The experiments are conducted using both the reduced-scale and full-scale protocols. In the reduced-scale protocol, the fusion process is performed in the downsampled or downscaled version of the PAN and MS images. Since the original MS images are intact, this evaluation denotes the full reference evaluation. The four universally accepted metrics of SAM [1], ERGAS [2], UIQI [3], and Q4 [4] are then computed. The optimal values for the global SAM and ERGAS are zero, and for UIQI and Q4 are one.

In addition to the above reduced-scale experiments, a full-scale resolution comparison is carried out. Since the reference MS image is not available at the PAN image scale for applying the full reference metrics, the no-reference quality metrics are employed. These metrics are D_s, D_λ, and QNR [5] as mentioned earlier. The optimal values for D_s and D_λ are zero while the reference value for QNR is 1.

The following conventional methods as well as deep learning-based methods are considered for comparison purposes: BDSD [6], Two-Step Sparse Coding (TSSC) [7], PNN [8], pansharpening with DNN [9], PanNet [10], Deep Residual Pansharpening Neural Networks (DRPNN) [11], and multi-objective deep learning based method (called MO here) [12], and unsupervised generative model for pansharpening (abbreviated as UGAN here) [13].

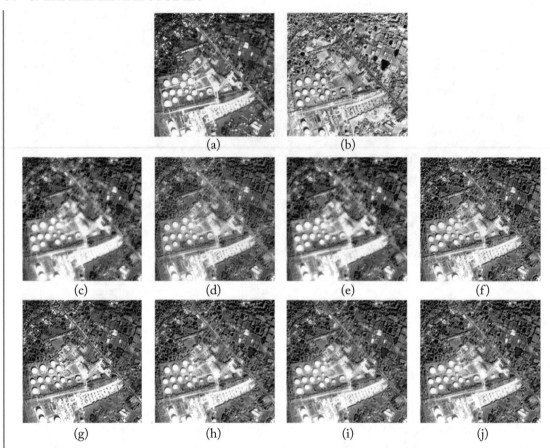

Figure 6.4: Reduced resolution sample fused images from the QuickBird dataset: (a) MS, (b) PAN at MS resolution, (c) BDSD, (d) TSSC, (e) PNN, (f) DNN, (g) PanNet, (h) DRPNN, (i) MO, and (j) UGAN.

6.3 VISUAL ASSESSMENT OF FUSION RESULTS

Besides the objective evaluation metrics reported in Tables 6.1 and 6.2, a sample fused image is provided here for visual comparison. For the full reference case, Figure 6.4 shows sample images from the QuickBird dataset. The PAN image in this figure is shown at the scale of the MS image. As can be seen from Figures 6.4c–6.4e, high spectral distortions are present. The light green color is completely changed to dark green for almost all the regions. Moreover, the color of the road crossing the green region is changed in these figures. In Figures 6.4f–6.4h, the level of injection of spatial details is high resulting in oversharpening. The fused images observed in Figures 6.4g and 6.4h suffer from the spectral distortion on the right side of the fused outcome.

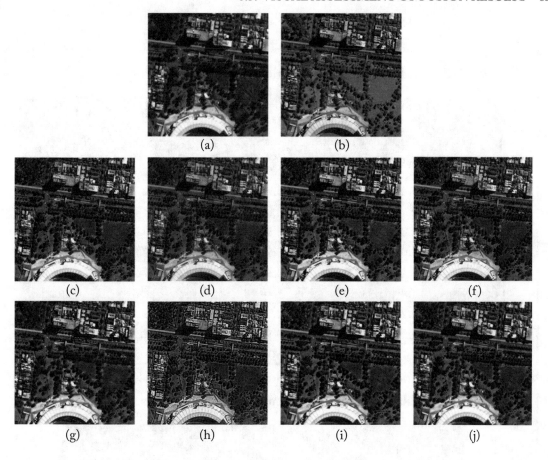

Figure 6.5: Reduced resolution sample fused images from the Pleiades-1A dataset: (a) MS, (b) PAN at MS resolution, (c) BDSD, (d) TSSC, (e) PNN, (f) DNN, (g) PanNet, (h) DRPNN, (i) MO, and (j) UGAN.

The fused outcomes of DNN, PanNet, and DRPNN appear similar to the MO and UGAN methods in some regions. However, in terms of the objective ERGAS and Q4 metrics (see Table 6.1), the UGAN method performs the best.

For the full-scale case, the fused images shown in Figure 6.8 can be visually compared. Since the resolution is high, in order to have an accurate comparison, a region of the image is cropped and placed in this figure. Figures 6.8c–6.8d, and 6.8f provide the blurry results in which the level of injection of spatial details is not sufficient. The spectral distortions in Figures 6.8e–6.8g appear high which result in color changes in some regions. The color distortion can be clearly seen in Figures 6.8g and 6.8h. The right region in Figures 6.8e and 6.8f exhibiting an intersection appears blurred. Figure 6.8g exhibits the fused outcome of the PanNet method

Table 6.1: Average reduced resolution objective metrics for three datasets

Methods	QuickBird				Pleiades-1A				GeoEye-1			
	SAM	ERGAS	UIQI	Q4	SAM	ERGAS	UIQI	Q4	SAM	ERGAS	UIQI	Q4
BDSD	2.26	1.45	0.97	0.84	5.20	4.88	0.89	0.86	2.72	2.02	0.95	0.86
TSSC	2.13	1.50	0.98	0.85	3.82	3.78	0.91	0.91	3.62	2.01	0.95	0.86
PNN	2.25	1.46	0.97	0.85	3.58	3.62	0.92	0.92	3.11	1.82	0.95	0.88
DNN	2.05	1.40	0.98	0.89	3.53	3.60	0.92	0.93	2.64	1.90	0.95	0.88
PanNet	2.13	1.50	0.97	0.88	4.58	4.02	0.91	0.91	2.23	1.91	0.95	0.87
DRPNN	1.90	1.45	0.98	0.89	3.88	3.67	0.91	0.92	2.49	1.82	0.96	0.87
MO	2.20	1.34	0.98	0.90	3.74	3.40	0.95	0.93	2.51	1.74	0.96	0.88
UGAN	**1.35**	**1.11**	**0.98**	**0.93**	**3.55**	**3.15**	**0.96**	**0.95**	**2.20**	**1.44**	**0.98**	**0.91**
IDEAL	0	0	1	1	0	0	1	1	0	0	1	1

Table 6.2: Average full resolution objective metrics for three datasets

	QuickBird			Pleiades-1A			GeoEye-1		
	D_s	D_λ	QNR	D_s	D_λ	QNR	D_s	D_λ	QNR
BDSD	0.05	0.04	0.91	0.06	0.04	0.90	0.08	0.05	0.87
TSSC	0.02	0.03	0.95	0.04	0.04	0.92	0.07	0.05	0.88
PNN	0.03	0.02	0.95	0.04	0.03	0.93	0.06	0.04	0.90
DNN	0.03	0.03	0.94	0.03	0.04	0.93	0.06	0.04	0.90
PanNet	0.03	0.04	0.93	0.04	0.02	0.94	0.05	0.05	0.90
DRPNN	0.03	0.03	0.94	0.04	0.02	0.94	0.05	0.04	0.91
MO	0.01	0.02	0.97	0.03	0.02	0.95	0.05	0.04	0.91
UGAN	**0.01**	**0.01**	**0.98**	**0.02**	**0.02**	**0.96**	**0.04**	**0.02**	**0.94**
IDEAL	0	0	1	0	0	1	0	0	1

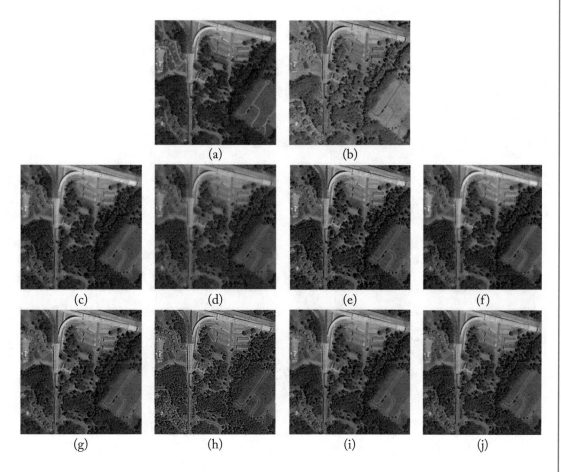

Figure 6.6: Reduced resolution sample fused images from the Geo-Eye1 dataset: (a) MS, (b) PAN at MS resolution, (c) BDSD, (d) TSSC, (e) PNN, (f) DNN, (g) PanNet, (h) DRPNN, (i) MO, and (j) UGAN.

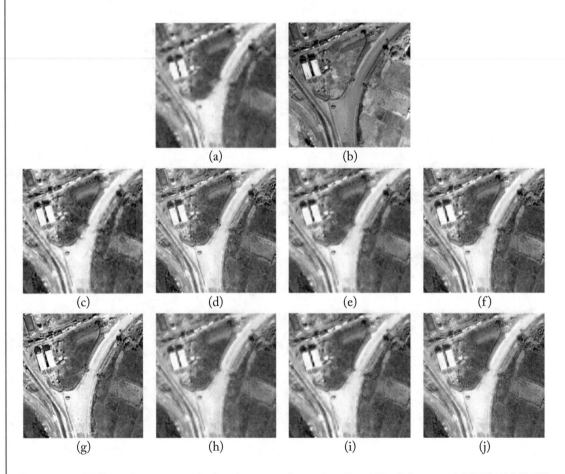

Figure 6.7: Full resolution sample fused images from the QuickBird dataset: (a) MS, (b) PAN at MS resolution, (c) BDSD, (d) TSSC, (e) PNN, (f) DNN, (g) PanNet, (h) DRPNN, (i) MO, and (j) UGAN.

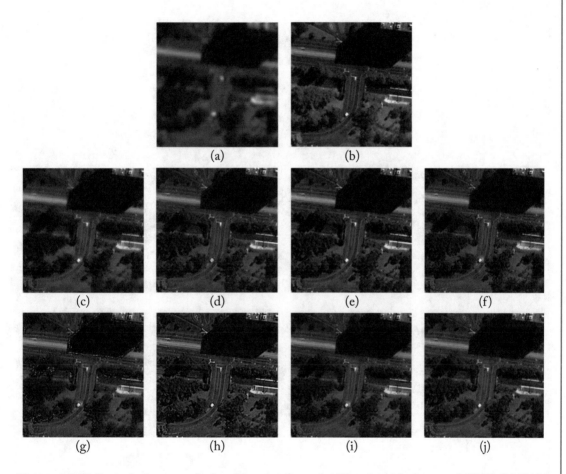

Figure 6.8: Full resolution sample fused images from the Pleiades-1A dataset: (a) MS, (b) PAN at MS resolution, (c) BDSD, (d) TSSC, (e) PNN, (f) DNN, (g) PanNet, (h) DRPNN, (i) MO, and (j) UGAN.

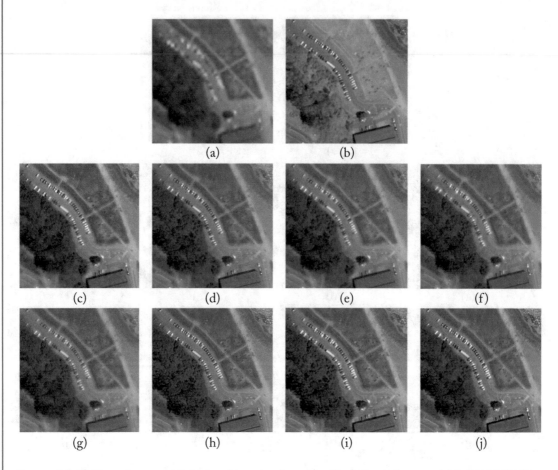

Figure 6.9: Full resolution sample fused images from the Geo-Eye1 dataset: (a) MS, (b) PAN at MS resolution, (c) BDSD, (d) TSSC, (e) PNN, (f) DNN, (g) PanNet, (h) DRPNN, (i) MO, and (j) UGAN.

suffering from color distortion. The fused outcomes of the multi-objective and UGAN methods are seen to be similar. From a visual perspective, the UGAN method shown in Figure 6.8j preserves the spectral content better than Figure 6.8i while injecting the spatial information into the fused image.

6.4 REFERENCES

[1] Yuhas, R. H., Goetz, A. F., and Boardman, J. W. 1992. Discrimination among semi-arid landscape endmembers using the spectral angle mapper (SAM) algorithm. 63

[2] Wald, L. 2000. Quality of high resolution synthesised images: Is there a simple criterion? 63

[3] Wang, Z. and Bovik, A. C. 2002. A universal image quality index. *IEEE Signal Processing Letters*, 9(3):81–84. DOI: 10.1109/97.995823. 63

[4] Alparone, L., Baronti, S., Garzelli, A., and Nencini, F. 2004. A global quality measurement of pan-sharpened multispectral imagery. *IEEE Geoscience and Remote Sensing Letters*, 1(4):313–317. DOI: 10.1109/lgrs.2004.836784. 63

[5] Alparone, L., Aiazzi, B., Baronti, S., Garzelli, A., Nencini, F., and Selva, M. 2008. Multispectral and panchromatic data fusion assessment without reference. *Photogrammetric Engineering and Remote Sensing*, 74(2):193–200. DOI: 10.14358/pers.74.2.193. 63

[6] Garzelli, A., Nencini, F., and Capobianco, L. 2007. Optimal MMSE pan-sharpening of very high resolution multispectral images. *IEEE Transactions on Geoscience and Remote Sensing*, 46(1):228–236. DOI: 10.1109/tgrs.2007.907604. 63

[7] Jiang, C., Zhang, H., Shen, H., and Zhang, L. 2013. Two-step sparse coding for the pansharpening of remote sensing images. *IEEE Journal of Selected Topics in Applied Earth Observations and Remote Sensing*, 7(5):1792–1805. DOI: 10.1109/jstars.2013.2283236. 63

[8] Masi, G., Cozzolino, D., Verdoliva, L., and Scarpa, G. 2016. Pan-sharpening by convolutional neural networks. *Remote Sensing*, 8(7):594. DOI: 10.3390/rs8070594. 63

[9] Huang, W., Xiao, L., Wei, Z., Liu, H., and Tang, S. 2015. A new pan-sharpening method with deep neural networks. *IEEE Geoscience and Remote Sensing Letters*, 12(5):1037–1041. DOI: 10.1109/lgrs.2014.2376034. 63

[10] Yang, J., Fu, X., Hu, Y., Huang, Y., Ding, X., and Paisley, J. 2017. PanNet: A deep network architecture for pan-sharpening. *Proc. of the IEEE International Conference on Computer Vision*, pages 5449–5457. DOI: 10.1109/iccv.2017.193. 63

[11] Wei, Y., Yuan, Q., Shen, H., and Zhang, L. 2017. Boosting the accuracy of multispectral image pan-sharpening by learning a deep residual network. *IEEE Geoscience and Remote Sensing Letters*, 14(10):1795–1799. DOI: 10.1109/lgrs.2017.2736020. 63

[12] Azarang, A. and Kehtarnavaz, N. 2020. Image fusion in remote sensing by multi-objective deep learning. *International Journal of Remote Sensing*, 41(24):9507–9524. DOI: 10.1080/01431161.2020.1800126. 63

[13] Azarang, A. and Kehtarnavaz, N. Unsupervised multispectral image fusion using generative adversarial networks. 63

CHAPTER 7

Anticipated Future Trend

In this final chapter, it is useful to state possible future research directions in multispectral image fusion as well as the key points covered in this lecture series book. In addition, some remaining unanswered questions are raised. As discussed in the previous chapters, deep learning models have been shown to perform more effectively compared to the conventional methods both in terms of spectral and spatial consistency. In most cases, the conventional methods consider the fusion problem as a globally minimized task which means the estimated parameters for injecting spatial details are acquired using the entire image rather than by considering regions-of-interest or local-based parameters. Even though some of the recently developed conventional methods have considered features in different regions instead of the entire image, the performance of deep learning models is still found to be superior.

The deep learning-based approaches to multispectral image fusion take into consideration the constraints associated with solving the ill-posed optimization problem of pansharpening. However, a major issue involves the computational time in which even a slight improvement in the fused outcome with respect to the state-of-the-art methods may be of computational significance. It is thus desired to develop time-efficient methods that can perform better than fast conventional methods. The increasing use of GPUs is envisioned to address the computational aspect.

The main drawback of the conventional methods is their inability to extract high-level features of the earth surface objects leading to non-optimal fusion outcomes. As noted in the previous chapters, the component substitution-based methods are known to generate poor fusion outcomes in terms of spectral consistency while the multi-resolution methods are known to generate poor fusion outcomes in terms of spatial consistency. Basically, future advancements in remote sensing fusion in terms of consistency between the spectral and spatial information can be made by addressing the following two major shortcomings of the conventional methods.

- In component substitution-based methods, the panchromatic data are replaced with an intensity component (I), which is histogram-matched with I. However, the matched version may contain spectral distortion and the same local radiometry, which leads to distortion in the fused outcome.

- Component substitution-based methods exhibit low sensitivity to spatial data while the multi-resolution analysis methods exhibit low sensitivity to spectral data.

The main challenges in both conventional and deep learning-based methods are as follows.

- Difference between the MS and PAN spectral response: In some cases, the earth observation satellites consist of four color channels, i.e., RGB and NIR, and the spectral signature of PAN data normally covers the entire range of MS data. A higher degree of overlap between these two signatures has a direct impact on the fused outcome. An issue to consider is that in some recently launched satellites, the color channels are wider and the MS data have up to eight bands. As a result, there exists less correlation between the PAN and MS spectral response. This issue is worth exploring in future studies.

- Mis-registration problem in urban regions: One critical issue in the fusion process is the PAN-MS misalignment, in particular, in urban areas when moving objects occur in the input data. This issue can diminish the performance of fusion for MRA-based methods. And in case of CS-based methods, although they are tolerant to mis-registrations to some extent, the fused outcome may have lower spectral fidelity. A focus of future studies can be placed on addressing this concern.

- High dimensionality of observations where MS and PAN images are often the input makes extracting spectral and spatial correlations of vital importance which is worth examining in future studies.

- Large amount of data encoded in each observation due to large distances between sensors and observation scenes, e.g., 200–400 km for low earth orbit satellites, creates significantly more data compared to a typical natural image. Future studies can look into mechanisms to cope with such large amount of data.

- Sensor specific radiometric resolution that unlike typical 8-bit imagery may involve 12-bits per pixel needs to be taken into consideration in future studies.

- Capturing reliable information which are adversely affected by environmental conditions, e.g., clouds or other type of weather conditions, is also an issue that requires further examination.

As noted in the previous chapters, several attempts have been made to solve the ill-posed problem of image fusion or pansharpening. The recent studies have revealed that the conventional loss functions do not generalize well.

- Most current deep learning models use stacked convolutional filtering processing layers to derive and fuse features from the input images. Such a stacked design does not completely tackle the flow and interaction of information. In addition, training challenges begin to rise as the network deepens. This element of the deep learning paradigm needs to be explored in future research.

- Current deep learning models generally optimize the network parameters by minimizing the MSE between the fused product and the corresponding reference data. Due to the absence of reference data, the spatially degraded reference images are often modeled when the original LRMS images are assumed to be reference images. Consequently, the training set contains down-scaled PAN and MS images and their corresponding reference MS image. This supervised learning technique is based on the assumption that a trained deep model at a lower-scale can be used on a coarser scale. While this issue has been shown to work in a number of cases, it can also cause scale-related challenges. This aspect is also worth looking at in future research.

- The mapping function defined by a DNN architecture may vary from sensor to sensor due to different properties/configurations of the optical sensing system. Consequently, the supervised learning process only yields a universal network model that may fail when applied to a sensing system whose training set has not been seen. This issue requires to be investigated in future studies.

- The mapping function extracts and injects details in a global fashion that may not reflect distinctive spectral characteristics in a local region. This issue is worth exploring in future studies.

Correction to: Image Fusion in Remote Sensing

Arian Azarang and Nasser Kehtarnavaz

Correction to:
A. Azarang and N. Kehtarnavaz, *Image Fusion in Remote Sensing*,
https://doi.org/10.1007/978-3-031-02256-2

The authors listed in the table of contents was incorrectly presented in the initially published online version of this book. The correct authors should read as Arian Azarang and Nasser Kehtarnavaz. This has now been corrected.

The updated version of this book can be found at
https://doi.org/10.1007/978-3-031-02256-2.

A. Azarang et al., *Image Fusion in Remote Sensing,*
© Springer Nature Switzerland AG 2023
DOI 10.1007/978-3-031-02256-2_8

Authors' Biographies

ARIAN AZARANG

Arian Azarang is a Ph.D. candidate in the Department of Electrical and Computer Engineering at the University of Texas at Dallas. His research interests include signal and image processing, machine/deep learning, remote sensing, and computer vision. He has authored or co-authored 15 publications in these areas.

NASSER KEHTARNAVAZ

Nasser Kehtarnavaz is an Erik Jonsson Distinguished Professor with the Department of Electrical and Computer Engineering and the Director of the Embedded Machine Learning Laboratory at the University of Texas at Dallas. His research interests include signal and image processing, machine/deep learning, and real-time implementation on embedded processors. He has authored or co-authored 10 books and more than 400 journal papers, conference papers, patents, manuals, and editorials in these areas. He is a Fellow of IEEE, a Fellow of SPIE, and a Licensed Professional Engineer. He is currently serving as Editor-in-Chief of *Journal of Real-Time Image Processing*.

Index

Printed in the United States
by Baker & Taylor Publisher Services